Applications of Advanced Nanomaterials

THE EDITOR

Dr. Tufan Ghosh received his M.Sc. in Chemistry from the Indian Institute of Technology Madras (IIT Madras). He completed his Ph.D. in Chemistry from IIT Madras under the supervision of Prof. Edamana Prasad. Later, he joined Prof. Sanford Ruhman's group at the Hebrew University of Jerusalem as a postdoctoral fellow and worked on understanding ultrafast charge carrier dynamics in various semiconducting bulk and nanomaterials. Currently, he is an assistant professor at the VIT-AP University, Amaravati, Andhra Pradesh where his group focuses on exploring charge carrier dynamics in halide perovskites. He is recipient of Start-Up Research grant from the DST-SERB, Govt. of India.

Dr. Roopas Kiran Sirugudu is currently working as an Associate Professor in the Department of Physics, School of Advanced Sciences at VIT-AP University. He received his PhD in Physics from Indian Institute of Technology Madras, India in 2012. He pursued his Postdoctoral Research work at University of Padua by associating with National Institute of Nuclear Physics (INFN, Legnaro), Italy from 2012-2014. He has more than 25 publications in various international journals in the field of ferroelectric and dielectric materials.

Dr. Sudagar Jothi is currently working as an Associate Professor in the Department of Physics, at VIT-AP University. He completed his PhD from the School of Materials Science and Engineering at Jilin University, China (2008-2012) with International scholarship between the MHRD, India and the MOE, China. He was a Visiting Research Associate at the Queen's University Belfast, UK in 2011. He pursued his Postdoctoral research work in High Performance Ceramics Lab, Metallurgical and Materials Engineering at IIT Madras, India (2013-2017) and in XRD Lab. He is the recipient of many awards including the prestigious International Corrosion Council fellowship in 2011 at Perth, Australia. He has published 30+ SCI papers in the field of High-Entropy materials, Eco-friendly Coatings, Corrosion science of engineering materials, Nanomaterials synthesis for Energy storage, and Sensors.

Applications of Advanced Nanomaterials

— *Editor* —

Tufan Ghosh
Roopas Kiran Sirugudu
Sudagar Jothi

2024

Daya Publishing House®
A Division of
Astral International Pvt. Ltd.
New Delhi – 110 002

ISBN: 978-93-5461-995-3

Published by : **Daya Publishing House®**
A Division of
Astral International Pvt. Ltd.
– ISO 9001:2015 Certified Company –
4736/23, Ansari Road, Darya Ganj
New Delhi-110 002
Ph. 011-43549197
E-mail: info@astralint.com
Website: www.astralint.com

Digitally Printed at : **Replika Press Pvt. Ltd.**

Preface

Nanoscience is a rapidly evolving field where innovations and discoveries are being reported every day. The study of nanomaterials is important as they offer unique chemical, optical, electrical, magnetic, and mechanical properties. In addition to their unique properties, advanced characterization and new fabrication techniques have placed nanomaterials as central to multiple disciplines, including materials science, chemistry, physics, engineering, and medicine.In this book titled "Applications of Advanced Nanomaterials", we present a list of recent developments in the field of nanoscience and nanotechnology. This book will give an overview of some of the recent research progress in the field of nanomaterial sciences, including synthesis, characterization, and applications of nanomaterials in nanoelectronics, optoelectronics, photovoltaics, and catalysis.

We thank the authors who have contributed to this book. We are also grateful to the reviewers who have put an enormous amount of effort to evaluate the chapters and provided their valuable suggestions.

Dr. Tufan Ghosh
Dr. Roopas Kiran Sirugudu
Dr. Sudagar Jothi

Contents

Preface *v*

1. **The Influence of Organic Dye on the Electrical Properties of n-Ge Based Organic-on-Inorganic Semiconductor Heterostructures** 1

   D. Mallikarjuna, A. Ashok Kumar, V. Rajagopal Reddy, V. Janardhanam, Chel-Jong Cho

2. **Effect of Reducing Agent on the Formation of Surfactant Free Tungsten Disulphide Nanoflowers** 11

   Saranya Sasi, Christeena Thomas, Benazeera Beegum and Reshmi R

3. **Lifetimes of the Metastable $n^2D_{5/2}$ States of K-like and Rb-like Ions** 17

   Narendra Nath Dutta

4. **An Optical Fiber-Fresnel-Reflection Sensor to Monitor Low-Temperature Response of Binary and Ternaryaqueous Solutions** 23

   Mani Priyadarshini, Archana Rallapalli, Venkata Rajanikanth Machavaram, Akella Sivaramakrishna, Pachaiappan Arulmozhivarman

5. **Structural Stability of Halide Perovskites: An Overview** 31

   *Vanga Ravali and Tufan Ghosh**

6. **Enhanced Magnetoelectric Coupling in Alternating Layers of Ferromagnetic and Ferroelectric Layers** 41

   V. K. Verma, V. R. Singh and S. Jena

7. **A Study on Silver Doped 2D MOS_2 Hybrid Nanostructure for Optoelectronic Device Applications** 47

Christeena Thomas, Benazeera Beegum K A, Saranya Sasi, Asha A S, Reshmi R

8. **Recent Advancements in The Photodegradation of Rhodamine B dyes** 55

Keerthi Valsalan and Triveni Rajashekhar Mandlimath

9. **An Insight into The Synthetic Methods of Carbon Dots and Hetero Atom Doped Carbon Dots** 67

Reshmi M K Nair and Triveni Rajashekhar Mandlimath

10. **Formic Acid, A Convenient Hydrogencarrier, and its Dehydrogenation by Palladium** 79

N Sai Kumar, Adhikary, A

11. **A Study of Carbon Nanotube FETs with Four Different Gate Insulators** 87

Md. Akram Ahmad, Jitendra Kumar and Bhubon Mech

12. **Macromolecular Crowding and its Impact on Biomolecular Processes and in the Fields of Nanomaterials and Nanotechnology** 97

Majumdar, Bibhab Bandhu

13. **Enhanced Performance of Lead-Free $Masni_3$ Based Perovskite Solar Cells: Numerical Approach** 103

Bhumika. E, N. Hima Varsha, Megha Unni, Sudagar Jothi, Chandu DS

14. **Orientation in A Large-Area Conducting Polymer Thin-Films for Organic Field-Effect Transistors Fabricated by Ribbon-Shaped FTM** 109

Atul S.M. Tripathi

15. **Review on Perovskite Solar Cells: Materials and Stability** 115

Vissam Setti Mourya Teja, Chinmaya Kumar Swain, Arun Kumar and L. Sowjanya Pali

16. **Catalytic Efficacy of Terpyridine Complexes in Carbon Dioxide Reduction** 135

Venkata Srujana B., Dr. Rajarshi Sarkar

17. Rhodium Catalyzed Organic Transformations: S Recent Perspective **145**

Sindhu Malisetti, Aparna Pulibandla, Divya Challa, Karri Eswar Brahma Reddy, Shaik Manjirunnisa, Dr. Dripta De Joarder

Index **157**

Chapter 1

The Influence of Organic Dye on the Electrical Properties of n-Ge Based Organic-on-Inorganic Semiconductor Heterostructures

D. Mallikarjuna[1,*], A. Ashok Kumar[1,*], V. Rajagopal Reddy[2], V. Janardhanam[3], Chel-Jong Choi[3]

[1]Department of Physics, YSR Engineering College of YVU, Proddatur, YSR Dist. Andhra Pradesh, INDIA

[1]Department of Physics, YSR Engineering College of YVU, Proddatur, YSR Dist. Andhra Pradesh, INDIA

[2]Department of Physics, Sri Venkateswara University, Tirupati, Andhra Pradesh, INDIA

[3]Semiconductor Physics Research Center (SPRC), School of Semiconductor and Chemical Engineering, Jeonbuk National University, Jeonju 54896, Korea

**Corresponding author: malli.arjuna1@gmail.com, drashok.yvuce@gmail.com*

INTRODUCTION

As the scaling of Si-based technology faces critical issues, the germanium (Ge) is considered as an alternative material in next-generation complementary metal oxide semiconductor (CMOS) devices to push forward the limitations of Si-based technology platforms due to its symmetric and high carrier mobility than Si. But strong Fermi-level pinning in the vicinity of the valence band maximum of Ge semiconductor raises the effective electron Schottky barrier height (SBH) of metal contacts on n-Ge [1–4]. Therefore, forming quality metal/Ge interface and ohmic contacts with n-Ge remains an obstacle in realizing high performance Ge-based CMOS devices. Several approaches have been proposed to overcome this obstacle, such as inserting interfacial layers (eg. Al_2O_3, TiO_2, SiO_2, ZnO, Mgo, Ge_3N_4, TaN) between a metal and n-Ge [5-10]. Similarly, the introduction of an organic interlayer to the metal/Ge interface is another approach to modulate the Schottky barrier height. For instance, we demonstrate the presence of a sulforhodamine-G (SRG) interlayer in the Au-Ge interface leads to the modification of the effective barrier height of Au/n-type Ge Schottky contact. Recently, Methylene Blue ($C_{16}H_{18}N_3ClS$), Orange-G ($C_{16}H_{10}N_2O_7S_2Na_2$), perylene monoamide ($C_{28}H_{24}N_2O_2$) have been considered as a very promising conducting organic dyes for various applications in electronic devices due to its superior properties such as high conductivity, structural stability, optical transparency, and processability [5,11-13]. Sulforhodamine-G is an ambipolar fluorescent dye with $C_{25}H_{25}N_2NaO_7S_2$ molecular formula which found significant applications in organic electronics as an acceptor dye. The device structure for electrical characterisation and the chemical structure of sulforhodamine-G is shown in Fig. 1

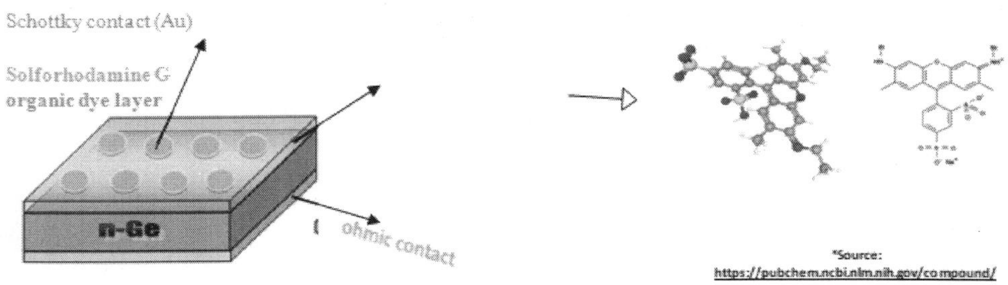

Fig. 1. Device structure for electrical characterization and Molecular structure (3D and 2D) of Sulforhodamine-G

Many researchers have studied about the influence of different interfacial layers on the performance of Au/n-Ge Schottky structures [14-18]. A low cost carbon paste thin film was coated on n-Ge and studied the electrical properties of the contacts. The use of carbon paste as an interlayer shows a highly stable and efficient contact and the device is found to operate even for 500h continuously without degrading its characteristics [14]. Kim et al [15] studies about the contact area dependent current transport mechanism of Au/n-Ge contacts. The Poole-Frenkel mechanism is more dominant for decreased contact area of the contacts. Yildirim fabricated Au/n-Ge contacts using Rubrene as interfacial layer [16]. The inclusion of Rubrene layer between Au and n-Ge shows improvement in device performance and enable high quality Schottky structures. Jyothi et al [17] studied about the electrical properties of Pt/pyronine-B/n-Ge heterostructure. The pyronine-B interlayer is coated on n-Ge using low cost spin coating method. The authors found improvement in device efficiency with pyronine-B interlayer and the current conduction mechanism at higher bias voltages are dominated by Poole-Frenkel emission due to the formation of defects or traps below the conduction band of n-Ge due to organic interlayer. A study on Schottky barrier properties of Au/n-Ge contacts with monolayer grapheme demonstrated the enhancement of contact parameters such as low ideality factor, low series resistance and higher barrier height than compared to Au/n-Ge conventional contacts [18]. The interfacial graphene layer not only passivates the semiconductor surface but also reduces the interface state density. Still the new materials show the pathway to understand the enhancement of electrical properties and current conduction mechanisms of various heterostructures using different interfacial layers.

In this work, we have fabricated Au/n-type Ge Schottky contact with a thin organic interlayer (SRG) using low-cost spin coating technique and demonstrated its electrical properties. Based on thermionic emission model, Schottky barrier parameters of Au/SRG/n-type Ge Schottky contact is evaluated using current–voltage (I–V) characteristics. The Cheung's and Norde methods are also utilised to evaluate the series resistance, barrier height and ideality factor to understand their consistency and validity. It was shown that SRG interlayer in-between metal and Ge substrate is effective in surface passivation of Ge and also influence the electrical properties of the conventional Au/n-Ge Schottky contacts.

MATERIALS AND METHODS

In this work, n-type Ge (100) (Sb-doped) wafers with a carrier concentration of 2 10^{18} cm^{3} is used to fabricate Au/SRG/n-Ge heterojunction. The wafer was degreased in trichloroethylene, acetone and methanol. Then, it was etched by H_2O:HF (100:1) solution to remove the native oxide from the surface of the wafer. Preceding each step, the wafer was rinsed in deionized water. Immediately after surface cleaning procedures, the sample was loaded into vacuum chamber to coat ohmic contact on the rough side of the sample with aluminum (Al) of 50 nm thickness. After ohmic contact formation, Sulforhodamine-G (SRG) film was coated with spin coater on the smooth side of the sample. Spin coater is operated at 4000 rpm with 60 sec duration to coat SRG on the smooth side of n-Ge surface and dried at 150 °C for 30 min. in ambient. Au thin film of 40 nm thickness was deposited using e-beam technique onto the top of the SRG thin film using a shadow metal mask. For a comparison, the Au Schottky contact without a SRG interlayer was fabricated on the n-type Ge wafer using same process conditions. The evaporation process was carried out at 5 10^{6} Torr. The electrical characteristics of the Au/SRG/n-Ge and Au/n-Ge Schottky contacts were measured at room temperature using a precision semiconductor parameter analyzer.

RESULTS AND DISCUSSION

Fig. 2 shows the I–V characteristics of Au/n-type Ge Schottky contact with and without SRG interlayer measured at room temperature. It is clear that both devices exhibited good rectifying properties. Based on thermionic emission model, the forward bias I–V characteristics of Schottky contact with the series resistance can be expressed as:

$$I_o = AA^{**}T^2 \exp\left(\frac{q\Phi_b}{KT}\right) \quad (1)$$

where q is the electron charge, V is the applied voltage, k is the Boltzmann constant, T is the absolute temperature, R_s is the series resistance, n is the ideality factor, I_o is the saturation current and can be obtained from the intercept of the plot of ln (I) versus V at V = 0 and the ideality factor 'n' can be determined from the slope of the linear region of the forward bias ln (I) vs. (V) graph expressed as:

$$I_o = AA^{**}T^2 \exp\left(\frac{q\Phi_b}{KT}\right) \quad (2)$$

where A is the area of the diode, A** is the Richardson constant (140 A cm^{-2} K^{-2} for n-type Ge) and $_b$ is the barrier height. Once the saturation current I_0 is determined, the barrier height $_b$ can be expressed as:

$$\Phi_b = \frac{kT}{q}\ln\left(\frac{AA^{**}T^2}{I_o}\right) \quad (3)$$

The barrier height and ideality factor were found to be 0.63 eV and 1.51 for the Au/n-type Ge Schottky contact, and 0.73 eV and 1.30 for the Au/SRG/n-Ge Schottky contact, respectively. Both devices exhibited an ideality factor deviating from the value of unity. This could be associated with the presence of a wide distribution of low-Schottky barrier patches caused by a laterally inhomogeneous barrier [19-20]. It is also evidenced that the barrier height of Au/SRG/n-Ge Schottky contact was higher than that of Au/n-type Ge Schottky contact. This indicates that the SRG interlayer effectively modulated the barrier height by influencing the interfacial potential barrier of the Au/Ge Schottky junction. The image-force effect, recombination–generation and tunnelling are all possible mechanisms that could lead to an ideality factor greater than unity. The Cheung's method [21] was employed to calculate the barrier height, ideality factor, and series resistance of Au/n-type Ge Schottky contact with and without SRG interlayer.

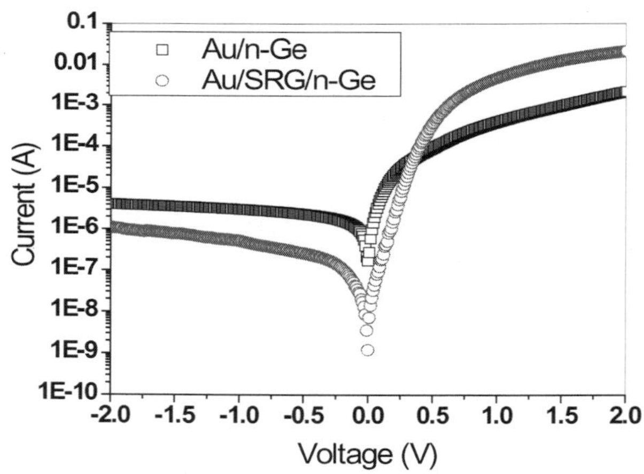

Fig. 2. I–V characteristics of Au/n-type Ge and Au/SRG/n-type Ge Schottky contact measured at room temperature.

Figure 3 shows a plot of dV/d(lnI) versus I and H(I) versus I for the Au/n-type Ge and Au/SRG/n-Ge Schottky contact. The series resistance and ideality factor were extracted from the slope and y-axis intercept of the plot respectively and obtained as 3027 and 1.97 for the Au/n-type Ge Schottky contact and 69 and 2.59 for the Au/SRG/n-type Ge Schottky contact. Using the value of ideality factor determined from the plot of dV/ d(lnI) versus I, the H(I) versus I plot gives a straight line with the y-axis intercept yielding the barrier height. The slope of H(I) versus I plot also provides a second method to determine the series resistance. The series resistance and barrier height extracted from the plot of H(I) versus I plot were found to be 2974 and 0.61 eV for the Au/n-type Ge Schottky contact, and 53 and 0.69 eV for the Au/SRG/n-type Ge Schottky contact, respectively. For both devices, the values of series resistance calculated from the plots of dV/d(lnI) versus I were almost identical to those from the plots of H(I) versus I.

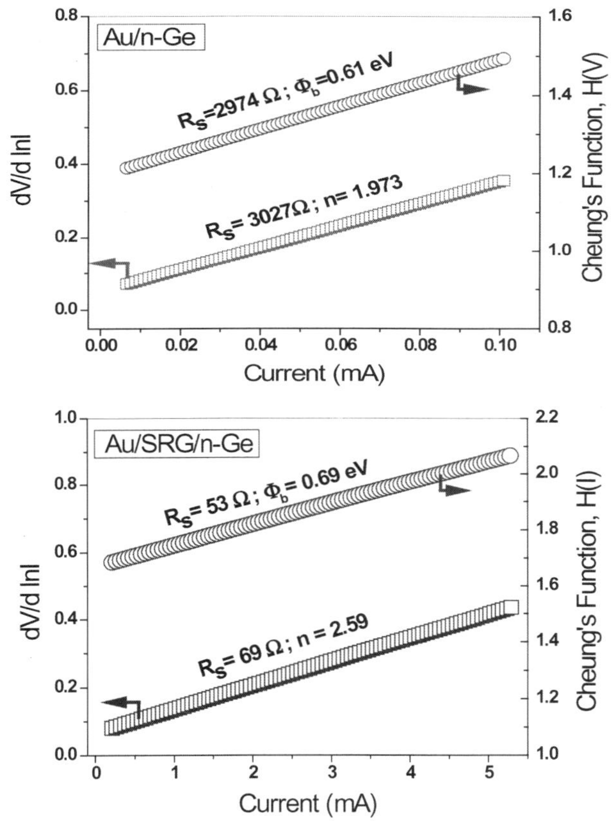

Fig. 3. Plot of dV/d(lnI) versus I and H(I) versus I of Au/n-Ge and Au/SRG/n-Ge Schottky contact.

Norde proposed an empirical function to calculate the barrier height and series resistance for a Schottky contact. The Norde function is defined as [22]:

$$F(V) = \frac{V}{\gamma} - \frac{kT}{q} \ln\left(\frac{I(V)}{AA^{**}T2}\right) \qquad (4)$$

where γ is an integer (dimensionless) greater than the ideality factor and I(V) is the current obtained from the forward bias I–V characteristics of the contacts. The basic technique of Norde method is to plot the proposed function with respect to the voltage applied across the Schottky contact, and to obtain its minimum. Once the minimum of the F(V) versus V plot is determined, the value of barrier height can be obtained using the following equation:

$$\Phi_b = F(V_{min}) + V_{min} - \frac{kT}{q} \quad (5)$$

where $F(V_{min})$, V_{min}, and I_{min} are the minimum value of F(V), the corresponding voltage and current, respectively. From the Norde function, the value of series resistance can be determined using the following equation:

$$R_s = \frac{kT}{q}\frac{(\gamma - n)}{I_{min}} \quad (6)$$

Figure 4 shows the plot of F(V) versus V of Au/n-type Ge Schottky contact with and without SRG interlayer. The barrier height and series resistance, extracted from the Norde plot, were found to be 0.63eV and 2458 Ω for the Au/n-type Ge Schottky contact, and 0.75eV and 73 Ω for Au/SRG/n-Ge Schottky contact, respectively. As shown in Table 1, a good consistency is observed between the values of barrier height, ideality factor and series resistance of Au/n-type Ge Schottky contact obtained using different approaches.

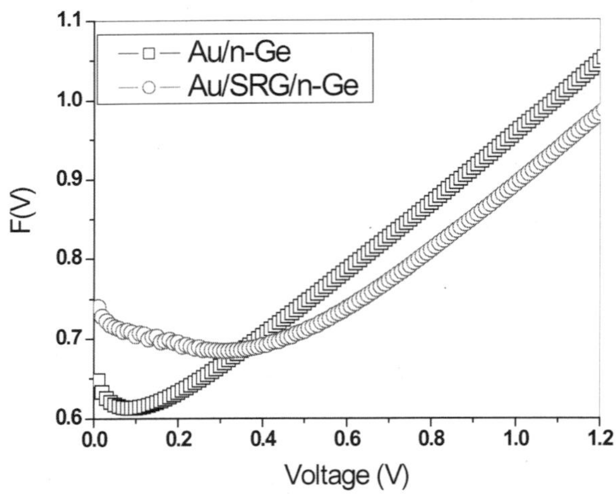

Fig. 4. Plot of F(V) versus V obtained from I–V characteristics of Au/n-type Ge and Au/SRG/n-type Ge Schottky contact.

Cheung's method and the Norde method, implying their validity for the evaluation of Schottky barrier parameters for the contacts with and without SRG interlayer. However, differences in the barrier height values obtained from these different techniques for the present devices could be attributed to the extraction parameters from different regions of the forward bias I–V plot [23-24]. For instance, ln (I) - V method and Cheung method are only applied to extract barrier height in the linear (low voltage) and non-linear (high voltage) regions of the forward bias I–V characteristics, respectively. Besides, the barrier height was calculated from the full forward bias I–V characteristics of the Schottky contact in Norde method.

The current conduction mechanisms of Au/n-Ge contacts with and without SRG interlayer was analysed using double logarithmic scale forward bias I-V characteristics. The double logarithmic scale I-V data is fitted linearly ($I\alpha V^m$) to identify the dominant current conduction at different bias regions for different magnitudes of slope factors ('m'). The observed dominant current conduction mechanism in Au/n-Ge contacts is due to ohmic conduction with slope factor m=1.59. on the other hand, three linear regions are observed in Au/SRG/n-Ge contacts. At low bias regions, the current-voltage follows a linear relation with slope (m=1.282) might be due to ohmic conduction but at moderately high voltage regions the slope is raised (m=5.546). The dominant current conduction mechanism observed is trap assisted space charge limited current in Au/SRG/n-Ge contacts at high bias regions. This might be due to the SRG interlayer introduces traps or defects at the interface below the conduction band of n-Ge semiconductor. Hence the current in these contacts are significantly controlled by traps or defects at the active layer of the Au/SRG/n-Ge contacts [16, 17]. Finally at high voltages, the evaluated slope factor found to be m=2.086, which might be due to trap-free space charge limited current conduction mechanism.

Table 1. Schottky barrier parameters extracted by using various methods of Au/n-type Ge and Au/SRG/n-type Ge Schottky contact.

Method	Sample	Ideality fact	Barrier heigh	Series resistanc
I-V graph	Au/n-Ge Au/SRG/n-Ge	1.51 1.30	0.63 eV 0.73 eV	— —
Cheung's Method (dV/dlnI) vs. I	Au/n-Ge Au/SRG/n-Ge	1.97 2.59	— —	3027 69
Cheung's Method (H(I) Vs. I)	Au/n-Ge Au/SRG/n-Ge	— —	0.61 eV 0.69 eV	2974 53
Norde Method	Au/n-Ge Au/SRG/n-Ge	— —	0.63 eV 0.75 eV	2458 73

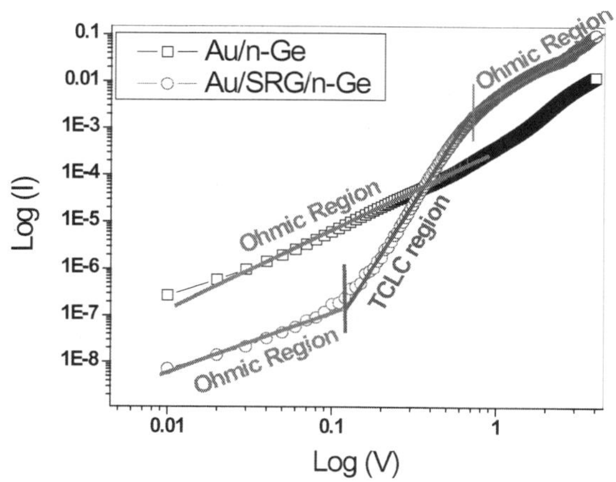

Fig. 5. Current conduction mechanism of Au/n-Ge and Au/SRG/n-Ge contacts in the Forward I-V characteristic.

CONCLUSION

In this work, we have investigated the electrical characteristics of the Au/SRG/n-Ge Schottky structures fabricated by spin coating of the organic dye material on n-Ge substrate. It has been seen that the SRG thin film on n-Ge substrate showed a good rectifying behaviour. The barrier height and the ideality factor of the device were calculated from the semi log I–V based on thermionic emission model. The Schottky barrier parameters such as barrier height, series resistance and ideality factor of the Au/SRG/n-Ge Schottky diodes compared with those of conventional Au/n-Ge contacts. The barrier height value of 0.73 eV obtained for the Au/SRG/n-Ge contact found significantly larger than the barrier height value of the conventional Au/n-Ge contact. Thus, the modification of the interfacial potential barrier for metal/Ge diodes has been achieved by using an SRG organic interlayer. This has been attributed to the fact that the SRG interlayer increased the effective $_{b}$ by influencing the space charge region of Ge. The insertion of organic dye (SRG) thin layer between Au/n-Ge contacts shows a trap assisted charge limited current at higher bias voltage due to the creation of defects or traps at the active layer of the contact below the conduction band.

REFERENCES

[1] MengraoTanga, Honghao Cai, The influence of ion-implantation on the effective Schottky barrier height of NiGe/n-Ge contacts, Chemical Physics 530 (2020) 110626

[2] JianyuanWang , Wei Huang, Jianfang Xu, Jun Li, Shihao Huang, Cheng Li, Songyan Chen, Schottky barrier height modulation effect on n-Ge with TaN contact, Materials Science in Semiconductor Processing 91 (2019) 206-2011

[3] V. Janardhanam, I. Jyothi, Jong-Hee Lee, Hyung-Joong Yun, Jonghan Won, Yong-Boo Lee, Sung-Nam Lee, Chel-Jong Choi, Electrical properties of a Cu-germanide Schottky contact to n-type Ge depending on its microstructural evolution driven by rapid thermal annealing, Thin Solid Films, 632 (2017) 23-27.

[4] Z. Wu, W. Huang, C. Li, H. Lai, S. Chen, Modulation of Schottky barrier height of metal/TaN/n-Ge junctions by varying TaN thickness, IEEE Tran. Electron Devices 59 (2012) 1328–1331.

[5] Y.S. Ocak, M. Kulakci, T. Kılıçoglu, R. Turan, K. Akkılıç, Current–voltage and capacitance–voltage characteristics of a Sn/Methylene Blue/p-Si Schottky diode, Synthetic Metals 159 (15–16) (2009) 1603-1607.

[6] Y. Zhou, M. Ogawa, X. Han, K.L. Wang, Alleviation of Fermi-level pinning effect on metal/germanium interface by insertion of an ultrathin aluminum oxide, Appl. Phys. Lett. 93 (2008) 202105-1 to 202105-3.

[7] J. Lin, A. Roy, A. Nainani, Y. Sun, K.C. Saraswat, Increase in current density for metal contacts to n-germanium by inserting interfacial layer to reduce Schottky barrier height, Appl. Phys. Lett. 98 (2011) 092113-1 to 092113-3.

[8] K. Martens, R. Rooyackers, A. Firrincieli, B. Vincent, R. Loo, B. De Jaeger, M. Meuris, P. Favia, H. Bender, B. Douhard, W. Vandervorst, E. Simoen, M. Jurczak, D. Wouters, J. Kittl, Contact resistivity and Fermi-level pinning in n-type Ge contacts with epitaxial Si-passivation, Appl. Phys. Lett. 98 (2011) 013504-1 to 013504-3.

[9] R. R. Lieten, S. Degroote, M. Kuijk, G. Borghs, Ohmic contact formation on n-type Ge, Appl. Phys. Lett. 92 (2008) 022106-1 to 022106-3.

[10] P.P. Manik, R. Mishra, V. Kishore, P. Ray, A. Nainani, Y. Huang, M. Abraham, U. Ganguly, S. Lodha, Fermi-level unpinning and low resistivity in contacts to n-type Ge with a thin ZnO interfacial layer, Appl. Phys. Lett. 101 (2012) 182105-1 to 182105-3.

[11] O. F. Yuksel, N. Tugluoglu, B. Gulveren, H. Safak, M. Kus, Electrical properties of Au/perylene-monoimide/p-Si Schottky diode, Journal of Alloys and Compounds 577 (2013) 30-36

[12] Ö. Güllü, A. Türüt, Electrical analysis of organic dye-based MIS Schottky contacts, Microelectronic Engineering, 87 (12) (2010) 2482-2487,

[13] R. Raji, K.S. Sibi, K.G. Gopchandran, ZnO:Ag nanorods as efficient photocatalysts: Sunlight driven photocatalytic degradation of sulforhodamine B, Applied Surface Science, 427 Part B, (2018) 863-875.

[14] P.T. Lin, J.W. Chang, S.R. Chang, Z.K. Li, W.Z. Chen, J.H. Huang, Y.Z. Ji, W.J. Hsueh, C.Y. Huang, A Stable and Efficient Pt/n-Type Ge Schottky Contact That Uses Low-Cost Carbon Paste Interlayers. Crystals 2021, 11, 259.

[15] Hogyoung Kim, Da Hye Lee and Hye Seon Myung, Contact Area-Dependent Electron Transport in Au/n-type Ge Schottky Junction. Korean J. Mater. Res. 26. (2016) 412-416.

[16] M. Yıldırım, Determination of Contact Parameters of Au/n-Ge Schottky Barrier Diode with Rubrene Interlayer, Journal of Polytechnic 20 (2017) 165-173

[17] I. Jyothi, V. Janardhanam, V. Rajagopal Reddy, Chel-Jong Choi, Modified electrical characteristics of Pt/n-type Ge Schottky diode with a pyronine-B interlayer, Superlattices and Microstructures, 75 (2014) 806-817.

[18] Zagarzusem Khurelbaatar, Yeon-HoKil, Hyung-Joong Yun, Kyu-Hwan Shim, Jung Tae Nam, Keun-Soo Kim, Sang-Kwon Lee, Chel-JongChoi, Modification of Schottky barrier properties of Au/n-type Ge Schottky barrier diode using monolayer graphene interlayer, Journal of alloys and compounds 614 (2014) 323-329.

[19] R.T. Tung, Electron transport at metal–semiconductor interfaces: general theory, Phys. Rev. B 45 (1992) 13509–13523.

[20] I. Jyothi, M.W. Seo, V. Janardhanam, K.H. Shim, Y.-B. Lee, K.-S. Ahn, C.-J. Choi, Temperature-dependent current–voltage characteristics of Er-silicide Schottky contacts to strained Si-on-insulator, J. Alloys Comp. 556 (2013) 252–258.

[21] S.K. Cheung, N.W. Cheung, Extraction of Schottky diode parameters from forward current–voltage characteristics, Appl. Phys. Lett. 49 (1986) 85–87.

[22] H. Norde, A modified forward I-V plot for Schottky diodes with high series resistance, Journal of Applied Physics 50 (1979) 5052-5053.

[22] M. Sharma, S.K. Tripathi, Study of barrier inhomogeneities in I–V–T and C–V–T characteristics of Al/Al_2O_3/PVA:n-ZnSe metal–oxide–semiconductor diode, J. Appl. Phys. 112 (2012) 024521-1–024521-10.

[23] Ö. Demircioglu, S. Karatas, N. Yildirim, Ö.F. Bakkaloglu, Effects of temperature on series resistance determination of electrodeposited Cr/n-Si/Au-Sb Schottky structure, Microelectron. Eng. 88 (2011) 2997–3002.

[24] A. Ashok Kumar, V. Rajagopal Reddy, V. Janardhanam, Hyun-Deok Yang, Hyung-Joong Yun, Chel-Jong Choi, Electrical properties of Pt/n-type Ge Schottky contact with PEDOT:PSS interlayer, Journal of Alloys and Compounds, 549 (2013) 18-21.

Chapter 2

Effect of Reducing Agent on the Formation of Surfactant Free Tungsten Disulphide Nanoflowers

Saranya Sasi[1], Christeena Thomas[1],Benazeera Beegum[1] and Reshmi R[1*]
[1]Optoelectronic and Nanomaterial's Research Laboratory, Department of Physics, Union Christian College, Aluva, Kerala-683102
E-mail (rreshmi@gmail.com)

INTRODUCTION

Tungsten disulphide (WS_2) is a representative of two-dimensional transition metal dichalcogenides *(TMDCs)*. The superior physical and chemical properties compared with other TMDCs tailor them for numerous applications including catalysis, energy storage devices, and sensors (Tian et al (2018), Huo et al (2014)). A metallic layer of tungsten (W) atoms was sandwiched between the top and bottom layers of sulphur (S) atoms producing an S-W-S layered structure (Yao et al (2015)). The strong covalent bond holds the atoms within a layer and the neighbouring layers were connected by the weak van der walls force (piao et al (2017)). Therefore the TMDCs can be exfoliated from bulk to mono/few layered structures. Such a scale down from bulk to few layers is accompanied by a gradual transition from indirect to direct band gap (Sasi et al (2022)). This transition features diverse properties that have not been observed in the bulk form (Gordon et al (2002), Mc Creary et al (2016)). Strong photoluminescence, high in plane carrier mobility and robust mechanical properties are some of the characteristics of mono/few layered structures (Yang et al (2014)). Owing to this unique layer dependent properties, there is a broad scope for 2D-TMDCs based applications like TFTs, photodetectors, biosensors etc (Sohail Ahmed & Jiabao Yi (2017), Huo et al (2014)).

Despite these factors, the synthesis of WS_2 nanostructures with uniform size and controlled morphology is still challenging. Several approaches have been adopted for preparing WS_2, including thermal evaporation, chemical vapour deposition (CVD) and atomic layer deposition (ALD) (Abbas et al (2020)). However, these techniques require high temperature environment and complex gas flow systems and ultimately high cost. The hydrothermal method is regarded as the low cost, simple and ecofriendly method for the synthesis of different nanostructures (Cao et al (2014)). There are many reports on the hydrothermal synthesis of a variety of nanostructures of WS_2. Cao etal synthesized flower like nanospheres of WS_2 using a surfactant assisted hydrothermal reaction (Cao et al (2014)). In another report, WS_2 nanorods were prepared by the hydrothermal reaction by Tang etal (2011).

In this work, we have reported the synthesis of flower like nanostructures of WS_2 by the hydrothermal reaction without any surfactant. The role of reducing agents in the formation of WS_2 nanoflowers was discussed.

MATERIALS AND METHODS

In the present work, WS_2 nanostructures were prepared by a simple hydrothermal reaction. Sodium tungstate (Na_2WO_4) and thioacetamide (CH_4N_2S) were used as the tungsten (W) and sulphur (S) sources respectively. We have tried two reducing agents, hydroxyl amine hydrochloride ($NH_2OH.HCl$) and oxalic acid ($C_2H_2O_4$). These three precursors were dissolved in distilled water in a 1:2:4 molar ratio respectively. The precursor solution was stirred for 1 hour and then transferred to teflon lined stainless steel autoclave. The hydrothermal reaction was carried out at a temperature of 200^0C for a period of 24 hours. After cooling, the precipitate was washed multiple times using water and dried at 200^0C in an oven for one hour to obtain WS_2 powder.

X-ray diffraction (XRD) spectrum of the prepared was recorded with Rigaku MiniFlex powder XRD having Cu-k radiation of λ=1.54A^0. The morphology of the samples was analysed using the Carl Zeiss field emission scanning electron microscope (FESEM) images. Raman spectrum of the sample was measured at room temperature by Lab RAM HR Spectrophotometer at an excitation wavelength of 514nm.

RESULTS AND DISCUSSION

Initially, sodium tungstate and thioacetamide was used as the tungsten and sulphur source respectively. The high solubility of sodium tungstate helps to carry out the hydrothermal reaction properly. In the absence of any reducing agent sodium tungstate and thioacetamide were combined together to form an intermediate yellow product, and WS_2 was not formed.

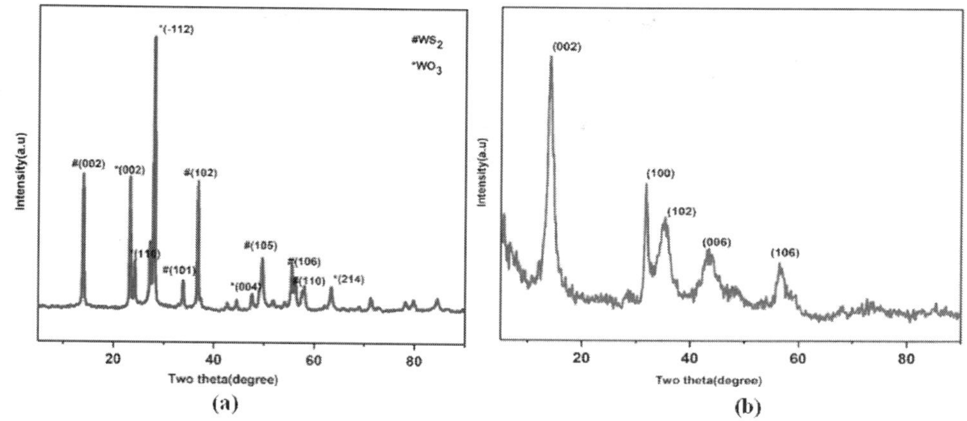

Fig:1 XRD pattern (a) WS_2-WO_3 composite and (b)WS_2

$$Na_2WO_4+CH_4N_2S+4H_2O\ 4CO_2+Na_2WS_4^{2-}+8NH_3 \text{ (Without Reducing agent)....(1)}$$

↓

[Intermediate complex]

So we have tried out two reducing agents **i)** Hydroxyl amine hydrochloride **ii)** Oxalic acid

Figure 1(a) shows the XRD pattern of WS_2-WO_3 composite using the hydroxyl amine hydrochloride as reducing agent. All the peaks were identified to both WS_2 and WO_3. WS_2exhibits the hexagonal phase (ICDD: **00-008-0237)** whereas WO_3 shows the monoclinic phase (ICDD:**00-043-1035)**. In this reaction, hydroxyl amine hydrochloride could partially reduce the intermediate complex $Na_2WS_4^{2-}$ and forms a WS_2-WO_3 composite instead of the complete formation of WS_2.

Figure1(b) depicts the XRD pattern of Pure WS_2, prepared using oxalic acid as the reducing agent. All the peaks were corresponding to the crystallographic planes of hexagonal WS_2 and no impurity peaks were detected. The strong peak at two theta value of 14.2^0 indexed to (002) plane indicates directional growth of well stacked layered structure of 2D WS_2 (Joseph et al (2021)). Here the strong oxalic acid was enough to reduce the intermediate complex completely into WS_2. This point out the formation of WS_2 occurs only in the presence of oxalic acid.

The FESEM images were used to identify the morphology of synthesized samples. The FESEM images of WS_2-WO_3 composite and pure WS_2 were displayed in fig:2. The composite shows an irregular arrangement of a large number of nanorods. Pure WS_2 prepared using oxalic acid exhibits flower like structures. A considerable quantity of WS_4^{2-} ions get reduced into WS_2 nuclei rapidly in the acid environment provided by the oxalic acid. These nuclei collectively begin to grow into larger crystals of WS_2. Abundant new nuclei are formed with the evolution of time at high temperature conditions. In the presence of oxalic acid, the nuclei of WS_2 nanosheets were gathered together to evolve into nanoflowers (Markus Niederberger and Helmut Colfen(2006)). The increased lattice energy upon the addition of oxalic acid causes the reorientation of the nucleus into the formation of sheet like structures of WS_2. Here the oxalic acid serves as a reducing agent and also the shape inducing factor in the formation of WS_2 nanoflowers (Keerthana et al (2021)).

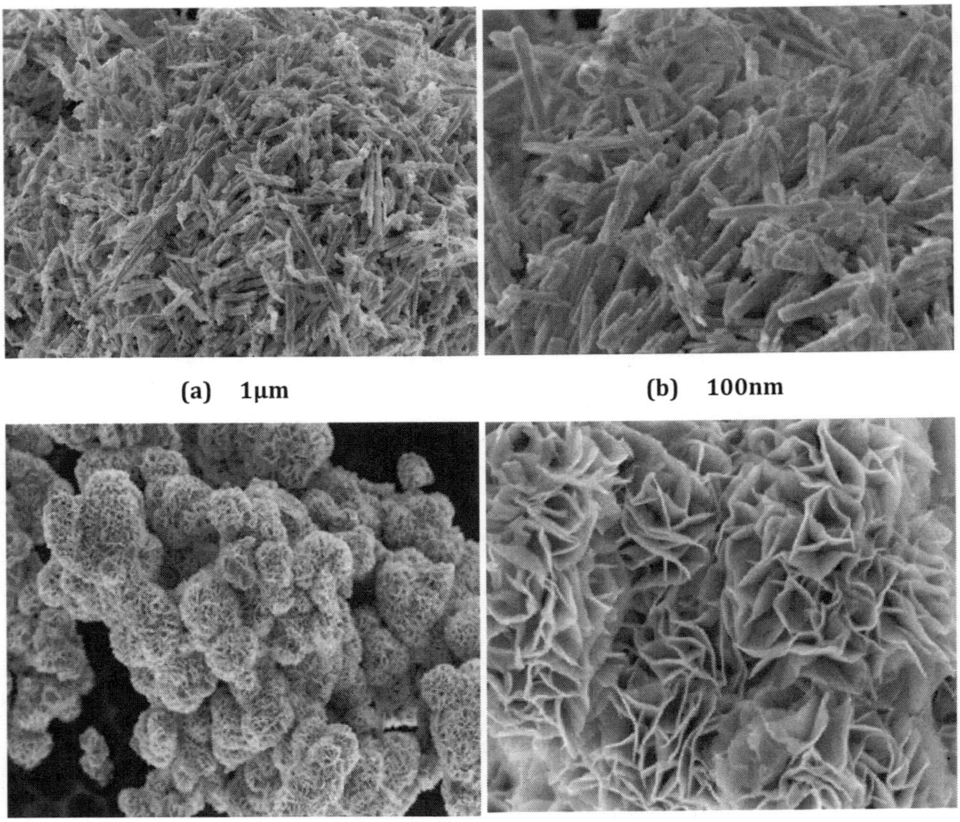

(a) 1μm (b) 100nm

(c) 1μm (d) 100nm

Fig:2 FESEM images of (a,b) WS_2-WO_3 composite and (c,d)WS_2

Fig: 3 shows the Raman spectrum of flower like WS_2 prepared by using oxalic acid as the reducing agent. It exhibits the presence of two characteristic peaks of WS_2. The first peak is located at 352.57cm^{-1} corresponding to the in plane vibration of atoms while the other one is at 418.65 cm^{-1} due to the out of plane vibration of the atoms (Loh et al (2015), Ali Eftekhari (2017)). There were no peaks were detected corresponding to WO_3 or any other impurities. The formation of pure WS_2 is evident from the Raman spectrum.

CONCLUSION

Flower like nanostructures of surfactant free WS_2 was prepared simple hydrothermal technique. The effect of reducing agents on the formation of WS_2 nanostructures was investigated. Oxalic acid is found to be essential for the formation of flower-like WS_2. Oxalic acid not only provides acidic medium as a reducing agent for the growth of WS_2, also serves as nucleation catalyst and shape inducing factor. Raman spectrum also confirms the formation of pure WS_2 in the presence of oxalic acid.

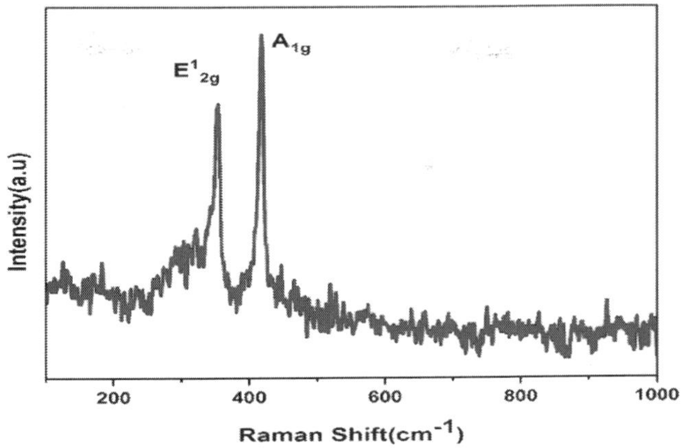

Fig. 3 Raman spectrum of WS_2 nanoflowers

ACKNOWLEDGMENT

The authors acknowledge DST-SERB (file No: CRG/2018/003785) for providing the financial assistance for this work.

REFERENCES

[1] Tian, K., Baskaran, K., & Tiwari, A. (2018). Growth of two-dimensional WS_2 thin films by pulsed laser deposition technique. Thin Solid Films, 668, 69-73.

[2] Huo, N., Yang, S., Wei, Z., Li, S. S., Xia, J. B., & Li, J. (2014). Photoresponsive and gas sensing field-effect transistors based on multilayer WS_2 nanoflakes. Scientific reports, 4(1), 1-9.

[3] Yao, J. D., Zheng, Z. Q., Shao, J. M., & Yang, G. W. (2015). Stable, highly-responsive and broadband photodetection based on large-area multilayered WS_2 films grown by pulsed-laser deposition. Nanoscale, 7(36), 14974-14981.

[4] Piao, M., Chu, J., Wang, X., Chi, Y., Zhang, H., Li, C.,& Joo, M. K. (2017). Hydrothermal synthesis of stable metallic 1T phase WS_2 nanosheets for thermoelectric application. Nanotechnology, 29(2), 025705.

[5] Sasi, S., Midhun, P. S., Joseph, A., Aneesh, P. M., Jayaraj, M. K., & Reshmi, R. (2022). Room temperature intrinsic ferromagnetism in pulsed laser ablated few layers of 2D-WS_2 on Si/SiO_2 substrates. Materials Today: Proceedings, 62, 5456-5459.

[6] Gordon, R. A., Yang, D., Crozier, E. D., Jiang, D. T., & Frindt, R. F. (2002). Structures of exfoliated single layers of WS_2, MoS_2, and $MoSe_2$ in aqueous suspension. Physical Review B, 65(12), 125407.

[7] McCreary, K. M., Hanbicki, A. T., Jernigan, G. G., Culbertson, J. C., & Jonker, B. T. (2016). Synthesis of large-area WS_2 monolayers with exceptional photoluminescence. Scientific reports, 6(1), 1-7.

[8] Yang, L., Cui, X, Zhang, J., Wang, K, Shen, M., Zeng, S., & Xiang, B. (2014). Lattice strain effects on the optical properties of MoS_2 nanosheets. Scientific reports, 4(1), 1-7.

[9] Ahmed, S., & Yi, J. (2017). Two-dimensional transition metal dichalcogenides and their charge carrier mobilities in field-effect transistors. Nano-micro letters, 9(4), 1-23.

[10] Huo, N., Yang, S., Wei, Z., Li, S. S., Xia, J. B., & Li, J. (2014). Photoresponsive and gas sensing field-effect transistors based on multilayer WS_2 nanoflakes. Scientific reports, 4(1), 1-9.

[11] Abbas, O. A., Zeimpekis, I., Wang, H., Lewis, A. H., Sessions, N. P., Ebert, M, & Sazio, P. (2020). Solution-based synthesis of few-layer WS_2 large area continuous films for electronic applications. Scientific reports, 10(1), 1-10.

[12] Cao, S., Liu, T., Hussain, S., Zeng, W., Peng, X., & Pan, F. (2014). Hydrothermal synthesis of variety low dimensional WS_2 nanostructures. Materials Letters, 129, 205-208.

[13] Cao, S., Liu, T., Zeng, W., Hussain, S., Peng, X., & Pan, F. (2014). Synthesis and characterization of flower-like WS_2 nanospheres via a facile hydrothermal route. Journal of Materials Science: Materials in Electronics, 25(10), 4300-4305.

[14] Tang, G., Tang, H., Li, C., Li, W., & Ji, X. (2011). Surfactant-assisted hydrothermal synthesis and characterization of WS_2 nanorods. Materials Letters, 65(23-24), 3457-3460.

[15] Joseph, A., Tadi, K. K., Anju, K. S., & Aneesh, P. M. (2021). Structural, optical, magnetic and electrochemical properties of hydrothermally synthesized WS_2 nanoflakes. Journal of Materials Research, 36(4), 884-895.

[16] Niederberger, M., & Cölfen, H. (2006). Oriented attachment and mesocrystals: Non-classical crystallization mechanisms based on nanoparticle assembly. Physical chemistry chemical physics, 8(28), 3271-3287.

[17] Keerthana, S., Rajapriya, A., Amirthapandian, S., Viswanathan, C., & Ponpandian, N. (2021). WS_2 hierarchical nanoflowers on rGO with enhanced electrochemical performance for sensitive and selective detection of mesalazine in real sample analysis. Colloids and Surfaces A: Physicochemical and Engineering Aspects, 618, 126452.

[18] Loh, T. A., & Chua, D. H. (2015). Origin of hybrid 1T-and 2H-WS_2 ultrathin layers by pulsed laser deposition. The Journal of Physical Chemistry C, 119(49), 27496-27504.

[19] AliEftekhari (2017). Tungsten dichalcogenides (WS_2, WSe_2, and WTe_2): materials chemistry and applications. Journal of Materials Chemistry A, 5(35), 18299-18325

Chapter 3

Lifetimes of the Metastable $n^2D_{5/2}$ States of K-like and Rb-like Ions

Narendra Nath Dutta
Department of Physics, School of Advanced Sciences,
VIT-AP University, Andhra Pradesh-522237, India
E-mail (narendranath.d@vitap.ac.in)

ABSTRACT

Here we represent the lifetimes of the first excited states of a few ions of K-isoelectronic and Rb-isoelectronic sequences. The lifetimes of these states are estimated using both theoretical and experimental parameters. The estimated lifetimes have values within a few seconds to a few hours. Also, the wavelengths of the transitions from these first excited states to the corresponding ground states fall in the mid-infrared region. Therefore, these ions may have potential applications as mid-infrared laser sources in the nano dimension.

INTRODUCTION

Theoretical and experimental studies on the long-lived or metastable states of ions can be of significant research interest for their relevance in stimulated emission of radiation. Such long-lived states can also have applications in precise spectroscopic measurements relevant to optical clocks and quantum information storage and processing [Safronova et al. (2014)]. There is an advantage of using ions with respect to neutral atomic species in high-precision spectroscopy. The electrons of an ion are strongly bound to the nucleus compared to the electrons of the neutral atom of the same electronic configuration. The binding force increases with the increasing charges of the ions. Consequently, the multiply charged ions are less affected compared to the atom by an external electric field and are shielded by the field-induced Stark shift to a greater extent. Therefore, the stimulated radiations emitted from these ions are supposed to be more monochromatic than the stimulated radiation emitted from the atom.

Here we represent the lifetimes of the metastable $n^2D_{5/2}$ states of a few ions belonging to the K-isoelectronic and Rb-isoelectronic sequences, where $n = 3$ and 4 for former and later sequences, respectively. The $n^2D_{5/2}$ states of these ions occupy the positions of the first excited states above the $n^2D_{3/2}$ ground states. The lifetimes of these excited states are calculated considering the electric quadrupole ($E2$) and magnetic dipole ($M1$) forbidden transitions [Bransden & Joachain (2006)] as the decay mechanisms of these states to the ground states [Dutta & Majumder (2016)]. Consequently, the lifetime of a particular ion depends on the $E2$ and $M1$ transition matrix elements between the $n^2D_{5/2}$ and $n^2D_{3/2}$ states and the transition wavelength between these two states [Dutta & Majumder (2016)]. These matrix elements are calculated here by using the highly correlated relativistic coupled-cluster method, and the transition wavelengths (experimental) are obtained from the atomic database of the National Institute of Standards and Technology (NIST) [Kramida et al. (2022)]. The detail of the computational procedure is given in Ref. [Dutta & Majumder (2016)]. The DIRAC-FOCK and the COUPLED-CLUSTER codes used for the computations were developed initially by the Indian Institute of Astrophysics (IIA), Bangalore, India group and modified later on by several, including the present author.

METHOD

The relativistic Hamiltonian for an atom or ion considered in our work has the following form [Clementi (1990)]:

$$H = \sum_i \left[c\alpha_i . p_i + (\beta_i - 1)c^2 - \frac{Z}{r_i} + \sum_{j<i} \left(\frac{1}{r_{ij}} - \frac{\alpha_i . \alpha_j}{2r_{ij}} - \frac{(\alpha_i . r_{ij})(\alpha_j . r_{ij})}{2r_{ij}^3} \right) \right], \quad (1)$$

where the summation signs on i and j indicate the electrons of the atom or ion. Here α_i and β are the Dirac Matrices. c is the velocity of light. Z is the nuclear charge of the atom or ion. r_i is the distance of 'i'th electron from the center of the nucleus. r_{ij} is the distance between 'i'th and 'j'th electrons. The first two terms on the right-hand side of the Eq. (1) represent the relativistic kinetic energies of the electrons. The third term represents the electrostatic energy due to the electron-nucleus Coulomb attractions. The fourth term indicates the energy due to the electron-electron Coulomb repulsions. The last two terms of the Eq. (1) represent the energy due to the electron-electron Breit interactions. The Breit interaction accounts the interaction between two Dirac currents [Mann & Johnson (1971)]. This form of the Hamiltonian can be used to solve the energy eigenvalue equation for a single-valence atom or ion having a valence electron at the 'v'th orbital:

$$H|\psi_v\rangle = E_v|\psi_v\rangle.$$

In coupled-cluster theory, the single valence state $|\psi_v\rangle$ is expressed in the form of cluster operators T and S_v operating on the reference state $|\varphi_v\rangle$ [Chaudhuri et al. (2003); Bartlett& Musial (2007)]:

$$|\psi_v\rangle = e^T \{1 + S_v\} |\psi_v\rangle.$$

Here the reference state $|\varphi_v>$ can be represented in the Slater determinant form using the independent particle model of the atomic Hamiltonian at the Dirac-Fock level [Clementi (1990)]. The operators T and S_v consider clusters of electron-excitations with respect to the reference $|\varphi_v>$. The wavefunctions $|\psi_v>$ are solved using the coupled-cluster theory for different single-valence states and are used to calculate the matrix elements of an operator ($\hat{O}$) using the expression [Dutta & Majumder (2016)]:

$$O_{fi} = \frac{\langle \psi_f | \hat{O} | \psi_i \rangle}{\sqrt{\langle \psi_f || \psi_f \rangle \langle \psi_i || \psi_i \rangle}}.$$

The matrix elements corresponding to the $E2$ and $M1$ operators can be employed to calculate the transition probabilities for the $E2$ and $M1$ transitions from the state $|\psi_f>$ to $|\psi_i>$ using the formulas [Dutta & Majumder (2016)]:

$$A_{E2}(f \rightarrow i) = \frac{1.1199 \times 10^{18}}{g_f \lambda^5} \left(O_{fi}^{E2}\right)^2,$$

and

$$A_{M1}(f \rightarrow i) = \frac{2.697 \times 10^{13}}{g_f \lambda^3} \left(O_{fi}^{M1}\right)^2,$$

respectively, in the unit of sec^{-1}. Here g_f is the degeneracy factor of the state $|\psi_f>$ and λ is the wavelength of the transition in the unit of Å. If the state $|\psi_f>$ has only one decay channel, $f \rightarrow i$,, through the forbidden transitions then the lifetime (τ_f) of this state can be calculated as the inverse of the total forbidden transition probability in the unit of sec:

$$\tau_f = \frac{1}{A_{E2}(f \rightarrow i) + A_{M1}(f \rightarrow i)}.$$

RESULTS AND DISCUSSION

Table I contains a first few ions of the K- and Rb-isoelectronic sequences corresponding to transition wavelengths in the mid-infrared region. The associated lifetimes of the $n^2D_{5/2}$states vary between a few hours to a few seconds, which can make these ions appropriate to achieve population inversion with respect to their ground state configurations. Here it is to be mentioned that the lifetimes of all the ions, except Cr^{5+}, were calculated without considering the Breit interaction. Whereas for Cr^{5+}, the lifetime was calculated using an unretarded approximation of the Breit interaction [Mann & Johnson (1971)]. Nevertheless, the contributions of the Breit interaction to the calculated lifetimes are expected to be little, and therefore their omissions do not affect the lifetimes of the ions within the significant digits. Also, the lifetime for Ti^{3+} is not considered exactly as reported in Ref.[Mandal et al. (2008)]. They considered the wavelength calculated by the relativistic coupled-cluster theory in their calculated lifetime. Table I presents the lifetime for this ion, which is calculated using the *E*2 and *M*1 matrix elements from their work and the wavelength as extracted from the NIST.

Table I: Lifetimes (in sec) of the $n^2D_{5/2}$ states with $n^2D_{5/2}$ - $n^2D_{3/2}$ transition wavelengths (in nm). n = 3 for K-isoelectronic and n = 4 for Rb-isoelectronic sequences.

Ions	Wavelengths	Lifetimes	Ref.
K-like ions			
Sc^{2+}	50597.05	12130.86	[Sahoo et al. (2008)]
Sc^{2+}	50597.05	12130.86	[Sahoo et al. (2008)]
Ti^{3+}	26171.16	1668.91	[Mandal et al. (2008)]
V^{4+}	16003.33	384.57	[Dixit et al. (2007)]
Cr^{5+}	10638.30	111.66	[Dutta & Majumder (2016)]
Rb-like ions			
Y^{2+}	13809.29	244.08	[Das et al. (2018)]
Zr^{3+}	7995.52	47.38	[Das et al. (2018)]
Nb^{4+}	5355.04	12.65	[Das et al. (2018)]
Mo^{5+}	3870.72	5.38	[Das et al. (2018)]

CONCLUSION

The ions from Sc^{2+} to Cr^{5+} in the K-isoelectronic sequence and Y^{2+} to Mo^{5+} in the Rb-isoelectronic sequence can be investigated as possible sources of mid-infrared laser light. Besides, the longer lifetimes of the first excited states may make these ions suitable to use in quantum technologies including quantum computation.

REFERENCES

[1] Bartlett, R. J., & Musial, M., 2007. Coupled-cluster theory in quantum chemistry. Rev. Mod. Phys., 79: 291.

[2] Bransden, B. H., & Joachain, C. J., 2006. Physics of Atoms and Molecules. Pearson Education.

[3] Chaudhuri, R. K. et al., 2003. Relativistic coupled cluster calculations of the energies for rubidium and cesium atoms. J. Chem. Phys., 119: 10633.

[4] Clementi, E. (Ed.), 1990. Modern Techniques in Computational Chemistry: MOTECC™-90. ESCOM Science Publishers B. V..

[5] Das, A. et al., 2018. Electron-correlation study of Y III-Tc VII ions using a relativistic coupled-cluster theory. J. Phys. B: At. Mol. Opt. Phys., 51: 025001.

[6] Dixit, G. et al., 2007. *Ab initio* calculations of forbidden transition amplitudes and lifetimes of the low-lying states in V^{4+}. Phys. Rev. A, 76: 042505.

[7] Dutta, N. N., & Majumder, S., 2016. Ab initio calculations of spectroscopic properties of Cr^{5+} using coupled-cluster theory. Indian J. Phys., 90: 373.

[8] Kramida, A. et al., 2022. NIST Atomic Spectra Database, http://physics.nist.gov/asd [Online].

[9] Mandal, S. et al., 2008. Theoretical spectroscopic studies of the atomic transitions and lifetimes of low-lying states in Ti IV. J. Phys. B: At. Mol. Opt. Phys., 41: 055701.

[10] Mann, J. B., & Johnson, W. R., 1971. Breit Interaction in Multielectron Atoms. Phys. Rev. A, 4: 41.

[11] Safronova, M. S. et al., 2014. Highly Charged Ions for Atomic Clocks, Quantum Information, and Search for α variation. Phys. Rev. Lett., 113: 030801.

[12] Sahoo, B. K. et al., 2008. Theoretical estimation of lifetimes of the lowest metastable states in Sc III and Y III. J. Phys. B: At. Mol. Opt. Phys., 41: 055702.

Chapter 4

An Optical Fiber-Fresnel Reflection Sensor to Monitor Low Temperature Response of Binary and Ternaryaqueous Solutions

Mani Priyadarshini[1,] Archana Rallapalli[2,] Venkata Rajanikanth Machavaram[3*], Akella Sivaramakrishna[4,] Pachaiappan Arulmozhivarman[5]

[1]*Department of Physics, Indian Institute of Technology Delhi, Hauz Khas, New-Delhi, India*

[2]*School of Electronics Engineering, VIT University, Vellore, Tamil Nadu*

[3*]*School of Advanced Sciences, VIT-AP University, Amaravati, Andhra Pradesh*

[4]*School of Advanced Sciences, VIT University, Vellore, Tamil Nadu*

[5]*School of Electrical Engineering, VIT University, Vellore, Tamil Nadu*

E-mail: machavvr@aol.co.uk

ABSTRACT

The design and construction of an optical fibre Fresnel reflection-based sensor to monitor the low-temperature response of aqueous solutions is carried out. The solutions of sodium chloride and calcium chloride salts dissolved in deionised water are subjected to cooling from 30 C to -200 C using liquid nitrogenand they are subsequently allowed to heat up at laboratory ambient conditions. The responseof the fibre optic sensor is monitored for varying solution concentrations and different phase change events have been inferred during cooling and heating phases. Thus, the binary fluids of aqueous sodium chloride at a eutectic concentration of 24.74 wt% and that corresponding to29.7 wt% of calcium chloride solution exhibited simultaneous crystallization into sodium chloride dihydrate with ice,and calcium chloride hexahydrate with ice respectively. Continued cooling of the solutions led to the formation of eutectic mixture in both the solutions.The existence of metastable liquid phases is prominent at -200 °C for a few minutes before the solution solidified into a eutectic mixture. This can be attributed to the phenomenon of supercooling, the degree of which is influenced by the salt composition and the solution concentration. Finally, ternary aqueous solution of NaCl-$CaCl_2$-$H_2O$is studied for low-temperature phase transition events.

INTRODUCTION

There is a wide appreiation regarding the low- temperature dynamics of aqueous solutions that are complex and a broad interst is shown by researchers towarsd research in this domain. This research interest is triggered by the unique property that these solutions are capable of remaining in the liquid phase for certain time duration at very low temperatures. Understanding the phase transitions in aqueous solutions is helpful in studying the influence of deicing agents in winter road safety monitoring [1], food preservatino industry [2], qualiy control in organ preservation [3], to study the implact of icing on aircraft structures [4], life supportiveness on outer terrestrial planets [5] as prominent applcation sectors. Hence, it is of high importance to study the low-temperature thermodynamics of aqueous solutions. Amongst the existing methods [6-8], optical fibre-based sensors extend the capability of deployment in remote and vulnerable environments, an important basis for their utility in several reported areas [9-12]. The fibre-optic sensors which work upon the Fresnel reflection phenomenon measure the magnitude of the reflected optical signals as dictated by the refractive index of the medium which lies in contact to the egde of the optical fiber. The refractive index of the medium changes based on parameters of the medium such as temperature, pressure, chemical composition, exposure to electromagnetic radiation [13,14]. Also, recent works on fibre-optic sensors for the applications such as monitoring phase changes in liquid assemblages [15-18], process-monitoring of hgh temperature materials [19, 20], refractive index sensing [21] are worth a mention. In this work, we have designed and developed a Fresenel-reflection based fibre-optic sensor to monitor and record the phase transitions that occur while we subject binary ad ternary aqueous solutions of sodium chloride and calcium chloride to continuously cooling from room temperature to -200 °C.

MATERIALS AND METHODS

(a) Sample solution preparation

Three types of sample solutions are prepared as explained: (a) 24.74wt% of sodium chloride (NaCl) dissolved in deionized water; (b) 29.7 wt% of calcium chloride ($CaCl_2$) solution; (c) Mixture of a 1:1 ratio of 21.4 wt% NaCl solution and 28 wt% of $CaCl_2$ solution. These test solutions are contained in a stainless-steel vessel and cooled below the room temperature to -200 °C using liquid nitrogen as a cooling agent.

(b) Design and deployment of the Fresnel fiber sensor

The edge of a single mode optical fiber is cleaved to establish perpendicularity with its axis and this cleaved fibre edge functions as the sensing region. This Fresnel sensor is placed adjacent to a k-type thermocouple at a spacing of < 4mm to simultaneously monitor the reflected optical signals and continuous variation in temperature as each sample solution is cooled. These sensors are immersed into the vessel containing the test liquid and the cooling process gets initiated soon as liquid nitrogen is introduced around the sample vessel. The interaction of the sample solution with the Fresnel sensor modifies the Fresnel reflectivity at the interface of the cleaved fiber end and the sample being cooled. A lateral spacing of ~4 mm minimizes the temperature gradient between the thermocouple and the Fresnel sensor.

RESULTS AND DISCUSSION

(a) Response of aq. sodium chloride solution to cooling

As per the phase diagram of NaCl solution [5], 24.74 wt% the NaCl crystallizes into $NaCl.2H_2O$ which is followed by the crystallization of the remaining solution into ice, thus forming a eutectic mixture of sodium dihydrate crystals and ice. The separation of $NaCl.2H_2O$ crystals alters the Fresnel reflectivity of the sample solution and there is release of heat. This exothermic event decreases the refractive index whereas the crystallization event tends to increase the refractive index. These effects indicate the Fresnel signal variation from points 'B' to 'C' which is determined by the cooling rate. The continued cooling of the solution leads to fast extraction of the heat released, thus resulting in the formation of ice from the solution. Thus, the solution gets solidified into initial formation of $NaCl.2H_2O$ crystals followed by ice which is together termed as the eutectic mixture. This, eutectic formation is shown by the variation from point 'C' to point 'D'. Further reduction in temperature by cooling densifies the eutectic mixture. The response of the Fresnel sensor during cooling is depicted in Figure 1 (a, b). The slow heating of this mixture starts after the liquid nitrogen is completely evaporated from around the steel vessel. A gradual rise in Fresnel signal until -91 °C is observed above which the signal suffered irregular variations until the stable eutectic temperature of -22.4 °C.

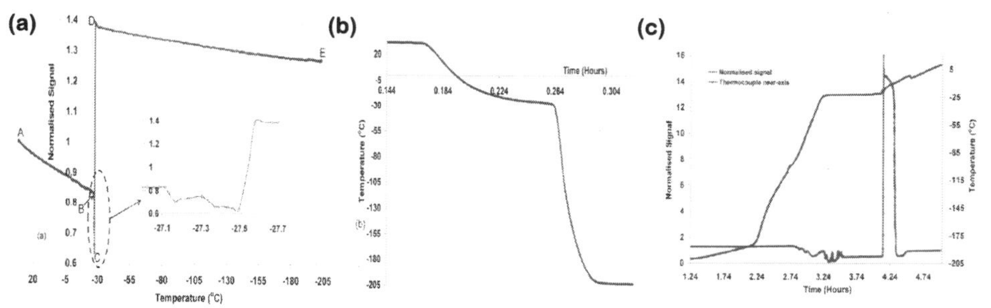

Figure 1 (a-b) Fresnel sensor response to the cooling of 24.74 wt% aq. NaCl solution (c) Sensor response for the room temperature heating of eutectic mixture

This implies that NaCl solution has undergone supercooling by about 6 to 8 °C below the eutectic temperature. By the end of eutectic phase, the eutectic mixture is completely transformed into solution form, after which the Fresnel signal trend followed that of temperature as shown in Figure 1c.

(b) Response of aq. calcium chloride solution to cooling

The phase diagram of aq. $CaCl_2$ [5] suggests that as the solution at eutectic concentration cools, phasechanges start to occur only as the temperature drops below the eutectic temperature. The trend of the thermocouple and the sensor signals is similar until -130 C, below which there is a slow reduction in the Fresnel signaluntil -200 C, Figure 2a shows this variation. It is observed that slope of the of the sensor response varies as a sixth-degree polynomialversus temperature with a local maximum at -63 C asthe temperature dropped below -50 C. This local maximum at-63 C is due to the effect of exothermic heat of crystallization of $CaCl_2$ $6H_2O$ (calcium chloride hexahydrate or antarcticite) which counteracts the cooling process.

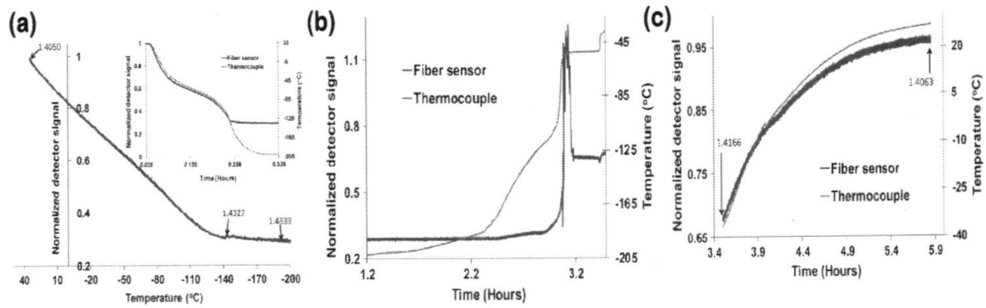

Figure 2. (a-b) Fresnel sensor response to the cooling of 29.7 wt% aq. $CaCl_2$ solution (c) Sensor response for the room temperature heating of the eutectic mixture

When the eutectic mixture begins to melt which occurred at -52 °C, there is rapid rise in the optical signal. A possible cause may be due to sensor debonding which causes significant back-reflection to get coupled into the fiber(see Fig. 2b). Eventually, the signal settled to a steady level through the melting phase. The Fresnel signal followed temperature trend continuously fashion as it increased towards the laboratory ambient. It can be inferred from the signal trend that the phase transformation of the eutectic mixture into solution occurs through simultaneous melting of ice and antarcticite during the heating phase (see Fig. 2c).

(c) Response of $NaCl$-$CaCl_2$-H_2O mixture to cooling

The concentration of the ternary solution corresponds to halite field of the phase diagram [22]. The sensor response in Figure 3a indicates a rapid decreasein temperature and the optical signals at ~-51 C which corresponds to the formation of eutectic mixture of ice (I), antarcticite(A, that is $CaCl_2.6H_2O$), and hydrohalite (HH, that is NaCl $2H_2O$). The most stablesensor response from -119 C indicates the completion of consolidationand densification of the eutectic mixture. A metastable liquid phase across the surface at -200 C is observed for few tens of minutes before the surface underwent solidification. The heating phase under laboratory ambience (see Fig. 3b) shows the influence of sensor debonding effect in the form of an abrupt rise and fall of signals, accompanied byoscillations in the signal. The subsequent near-stable temperature region between -55 C and -52.5 C corresponds to the first eutectic melting event. For the considered concentration of the ternary solution, there are two possibilities for melting of the ternary eutectic mixture at the temperature of-52 C.

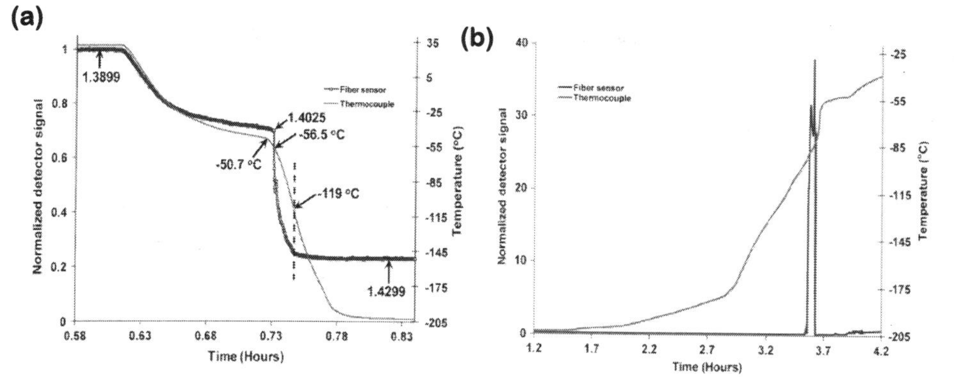

Figure 3. (a) Fresnel sensor response to the cooling of $NaCl-CaCl_2-H_2O$mixture (b) Sensor response for the room temperature heating of the eutectic mixture

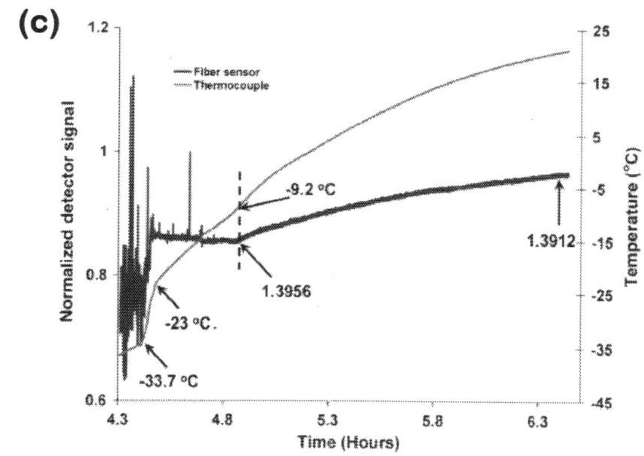

Figure 3. (c) Sensor response post the eutectic temperature for gradual heating of the sample mixture

The first possibility involves dissolutionof antarcticite followed by evolutionalong the cotectic of I and HH until the dissolution ofice. Subsequently, the sample mixture emerges along the HH field toward peritectic of HH and halite (H). The evolution of the medium continues to occur along the peritectic until there is transformation of HH to H (NaCl). Finally, the liquid evolves along a straight line towards the peak of NaCl to reach the original composition as the temperature increases. The second possibility involves dissolution of I at the ternary eutectic temperature followed by evolution along the cotectic of A and HH. Following thedissolution of A, the medium emerges through the HH fieldon to the peritectic of HH and H. The HH transforms into H along the peritectic and enters the H field. The dissolution of A is marked by the end of eutectic phase at -52 C as seen in Figure 3b. Thereafter, the sensor signals increased with temperature until -33.7 C (see Fig. 3c). Some variations in signal are due to the interaction of light with the dissolving ice at the end of the fiber. The rise in the normalized signal level to 0.87 at-33.7 C indicates decreased refractive index of the medium due to the dissolution of ice. The signal evolution from -23 C to-9.2 C may be due to transformation of HH to H. Subsequently, the signal followed the temperature trend resulting in original composition of the solution. These possibilities which lead the medium to halite field is similar to that reported by Bakker and Baumgartner [22].

CONCLUSION

A low-cost Fresnel reflection-based fibre sensor system is demonstrated for detecting the various phase change events during the cooling of aqueous solutions of NaCl, $CaCl_2$ and a ternary solution of NaCl-$CaCl_2$. The sensor responses to crystallization and dissolution events are found to be reproducible under the given experimental conditions. These sensors can aid in determining the optical properties of metastable phases at low temperatures and offer multipoint measurements from remote locations within the bulk of a relatively large volume of the medium subjected to cooling. It is envisaged that this low-cost sensor can offer attractive applications in bio-preservation, polymer processing and mineralogy.

ACKNOWLEDGEMENT

This work was carried out when the authors are based in VIT Vellore. The authors acknowledge the RGEMS grant provided by VIT University. VRM would like to thank Professor Gerard Fernando of University of Birmingham, UK. We are thankful to OPTICA, United States for permitting us to reproduce the contents of references 15 and 16

REFERENCES

[1] Terry, L.G., K. Conaway, J. Rebar et al., 2020. Alternative deicers for winter road maintenance, Water Air Soil Pollut. 231, 394

[2] Li, B. and D.W. Sun. 2002. Novel methods for rapid freezing and thawing of foods: a review, J. Food Eng. 54, 175–182

[3] Pegg D.E. and A.M. Karow. 1987. The Biophysics of Organ Cryopreservation (Springer, 1987)

[4] Gent, R.W., N.P. Dart and J.T. Cansdale. 2000. Aircraft icing, Philos. Trans. R. Soc. A 358, 2873–2911

[5] Toner, J. D., D.C. Catling, and B. Light. 2014. The formation of supercooled brines, viscous liquids, and low-temperature perchlorate glasses in aqueous solutions relevant to Mars, Icarus 233, 36–47

[6] Wilson, P.W., J.W. Arthur, and A.D. J. Haymet. 1999. Ice premelting during differential scanning calorimetry, Biophys. J. 77, 2850–2855

[7] Amamou, A.H., H. Benkhelifa, G. Alvarez, and D. Flick. 2010. Study of crystal size evolution by focused-beam reflectance measurement during the freezing of sucrose/water solutions in a scraped-surface heat exchanger, Process Biochem. 45, 1821–1825

[8] Borisenko, A.S. 1977. Study of the salt composition of solutions of gas-liquid inclusions in minerals by cryometric method, Geologija i Geofyzika 18, 16–27

[9] Pu, S., X. Chen, Y. Chen, W. Liao, L. Chen, and Y. Xia. 2005. Measurement of the refractive index of a magnetic fluid by the retroreflection on the fiber-optic end face, Appl. Phys. Lett. 86, 171904

[10] Avdeev, M.V., A. N. Konovalov, V. N. Bagratashvili, V. K. Popov, S. I. Tsypina, M. Sokolova, J. Ke, and M. Poliakoff. 2004. The fibre optic reflectometer: a new and simple probe for refractive index and phase separation measurements in gases, liquids and supercritical fluids, Phys. Chem. Chem. Phys. 6, 1258–1263

[11] Prada, D., C. Martelli, C.C. Kato, A.M.B. Braga, and M.S.P. Gomes. 2011. CO_2 phase study using an optical refractometer, Proc. SPIE7753, 775366

[12] Burton, G., L. Melo, S. Warwick, M. Jun, B. Bao, D. Sinton, and P. Wild. 2014. Fiber refractometer to detect and distinguish carbon dioxide and methane leakage in the deep ocean, Int. J. Greenhouse Gas Control 31, 41–47

[13] Zou, J., L. Ye, J. Ge, and C. Zhao. 2013. Novel fiber optic sensor for ice type detection," Measurement 46, 881–886

[14] Kim C.B. and C.B. Su. 2004. Measurements of the refractive index of liquids at 1.3 and 1.5 micron using a Fiber-optic Fresnel ratio meter, Meas. Sci. Tech. 15, 1683–1686

[15] Mani, P., A. Rallapalli, V.R. Machavaram, A. Sivaramakrishna. 2016. Monitoring phase changes in supercooled aqueous solutions using an optical fiber Fresnel reflection sensor, Opt. Exp. 24 (5), 5395-5410

[16] Priyadarshini, M., V.R. Machavaram, A. Sivaramakrishna and P. Arulmozhivarman. 2017. Detecting phase transitions in a $CaCl_2$-H_2O system at low temperatures using a fiber-optic Fresnel reflection sensor, Appl. Opt. 56 (11), 3229-3239

[17] Han, W., M. Rebow, D. Liu, G. Farrel, Y. Semenova and Q. Wu. 2018. Optical-fiber Fresnel reflection sensor for direct detection of the solid-liquid phase change in n-octadecane, Meas. Sci. Tech. 29 (12), 125107

[18] Wahl, M.S., A. Aasen, D.R. Hjelme and O. Wilhelmsen. 2020. Ice formation and growth in supercooled water-alcohol mixtures: Theory and experiments with dual fiber sensors, Fluid Phase Equilibira 522, 112741

[19] Zhao, X., Z. Zhang, K. Guo, J. He, G. Zhou and Z. Huang. 2020. Fresnel reflection-baed fiber sensor for UV adhesive cure monitoring," Opt. Comm. 474, 126099

[20] King, D. Gareth, Fiber-optic sensor development for process monitoring of epoxy resins," Univ. Birmingham (2018)

[21] Y. Wu, L. Xia and N. Cai, "Dual-wavelength intensity-modulated Fabry-Perot refractive index sensor driven by temperature fluctuation," Opt. Lett. 43 (17), 4200-4203 (2018)

[22] Bakker R.J. and M. Baumgartner. 2012. Unexpected phase assemblages in inclusions with ternary-H_2O–salt fluids at low temperatures, Central Eur. J. Geosci. 4, 225–237

Chapter 5

Structural Stability of Halide Perovskites: An Overview

Vanga Ravali and Tufan Ghosh*
Department of Chemistry, School of Advanced Sciences,
VIT-AP University, Amaravati,
Andhra Pradesh, 522237, India,
Email: tufan.ghosh@vitap.ac.in

ABSTRACT

Organic-inorganic lead halide perovskites are strong light-absorbing materials for low-costsolar energy generation. The power conversion efficiency (PCE) of certain halide perovskite devicesare reported to be greater than 25% in recent years. This excellent PCE comes from a strong absorption coefficient in the visible region of the solar spectrum, slow carrier recombination rate andlarge carrier diffusion length in such materials. Recently reports suggest that despite of being delivering high efficiencies, three-dimensional halide perovskite in particular when exposed to ambient atmosphere are unstable for long period. In this book chapter, we review of stability of 3D perovskites recently reported in the literature. Using analytical tools like UV-visible absorption and powder x-ray diffraction studies, the stability of the perovskite films can be monitored. We also discuss how reducing the dimensionality from three to two potentially increases the stability of the perovskite films, however, at a cost of reduced PCE.

INTRODUCTION

In recent years, researchers are showing more interest towards the solid-state light absorbing materials mainly on hybrid halide perovskites. Most of the recent studies are based on three dimensional (3D) methylammonium lead halide perovskite with representative formula $MAPbX_3$ (MA: methylammonium cation ($CH_3NH_3^+$), X: halide anions (Cl^-, Br^-, I^-)) (Fig. 1b) [1,2,3]. Organicinorganic hybrid perovskites $CH_3NH_3PbI_3$, in the form of dye sensitized solar cells first employed byMiyasaka and co-workers [4]. The unique properties of these high efficient materials are large absorption cross-section in the visible range, optimum band gap, slow charge carrier recombination rates and long diffusion length. Approximately a decade ago, Miyasaka and co-workers reported an efficiency of their perovskite solar cells less than 4% in the year 2009. In recent years, maximum efficiency of nearly 25% is reported by various research groups [5,6]. Although the efficiency reachescloser to the efficiency of silicon-based cells but the stability of the perovskite solar cells are reportedto be far less to that of silicon cells. Currently therefore one of the most concerning to the perovskiteresearch is the stability of these materials towards ambient atmosphere. Recent studies observed thatone of the most prominent reasons behind such unstable nature of the 3D perovskite could be the presence of the methylammonium (MA) groups in the 3D structure [7].

Lately, the focus in the field of perovskite research has shifted towards making stable perovskite materials even at the cost of sacrificing power conversion efficiency. To increase the stability, researchers replaced some of the MA groups with large organic groups like n-butylamine (BA) and phenylethylamine (PEA) in 3D perovskites [8]. Apparently, BA and PEA are hydrophobic in nature and due to the size difference between these organic groups to the inorganic groups there is a formation of two-dimensional (2D) layered structure depending upon their ratios. General representation of 2D perovskites is $(RNH_3)_2A_{n\,1}M_nX_{3n+1}$ (n = 1, 2, 3, 4.), where RNH_3 is a bulky

organic ammonium cation like BA and PEA, A is the monovalent cation typically methylamine (MA),M is a divalent metal cation (Pb^{2+}, Sn^{2+},..etc.), X is a halide and n is the number of $[MX_6]^{4-}$ layers within the quantum well (Fig. 1a) [9]. It has been observed that although the 2D perovskite devices are comparatively more stable than 3D perovskite films, however, the PCE is lower than that of the 3D perovskites [10].

3D PEROVSKITES

Recent studies reported the power conversion efficiency of these 3D perovskites about 25% which is significantly higher compared to the first perovskite cell with an efficiency about 4% reported in 2009 [4]. Moreover, in the last few years, continuous efforts are in place towards improving the efficiency of the 3D PSCs. Park and co-workers have reported an efficiency of 6% [11]. In 2012, it reaches 10-11% on solid-state perovskite solar cells reported by three groups coincidentally [12,13,14]. The major reasons behind such high efficiencies are: long carrier diffusion lengths, tunable band gap with great light absorption potential similar to that of Si-solar cells. However, most of the concerns related to this 3D perovskite materials is the fact that they are relatively much less stable compared to that of Si-cells [15].

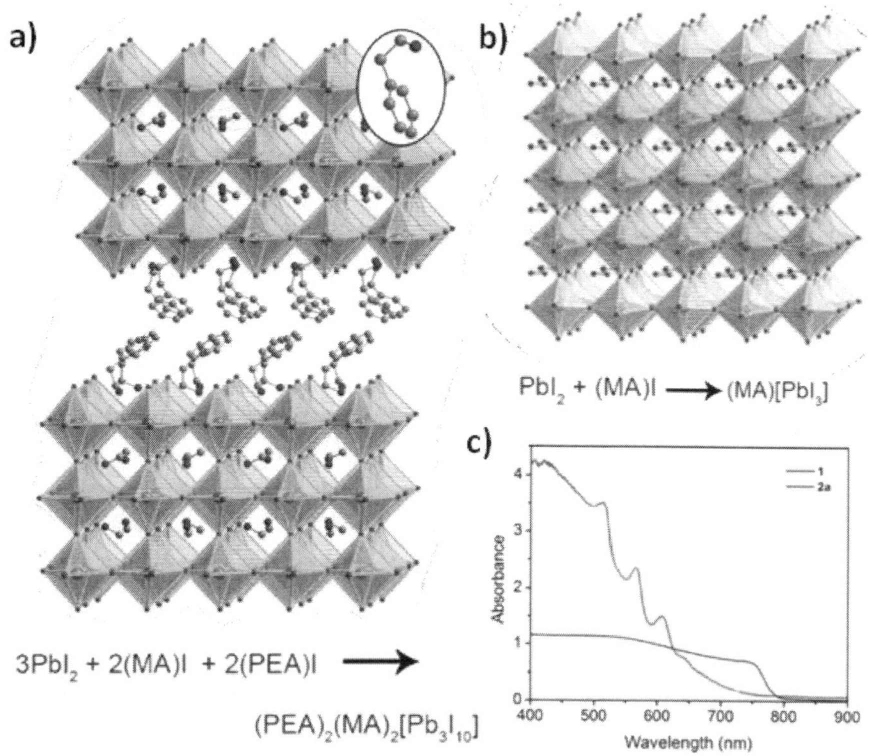

Fig. 1. Representative crystal structures of the (a) 2D perovskite $(PEA)_2(MA)_2[Pb_3I_{10}]$, (b) 3D perovskite $(MA)[PbI_3]$, and (c) Comparison of absorption spectra of $(PEA)_2(CH_3NH_3)_2[Pb_3I_{10}]$ (redsolid line) and $(MA)[PbI_3]$ (blue solid line). [taken from ref. 8]

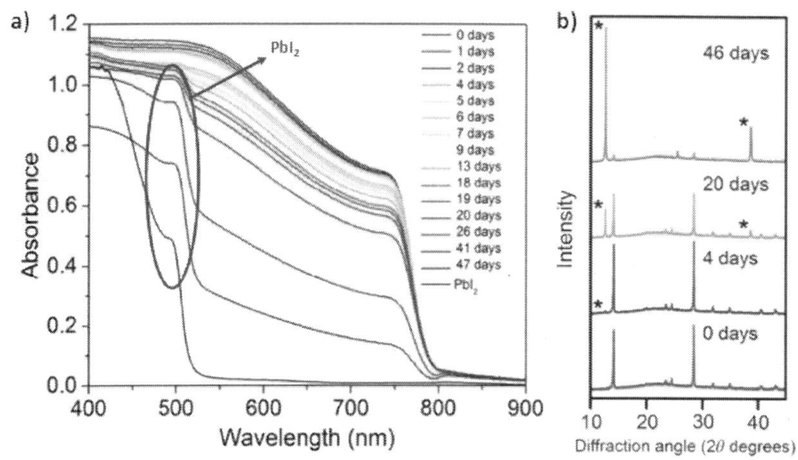

Fig. 2. (a) UV-visible absorption spectra, and (b) PXRD patters of $(MA)[PbI_3]$ upon exposure to 52% relative humidity at various time intervals. [taken from ref. 8]

The degradation of the MAPbI3 films can be monitored using various techniques including UV- visible absorption measurements, powder X-ray diffraction (PXRD), transient absorption studies to name a few. The absorbance measurements for the 3D perovskites MAPbI3 films exhibit at visible range with two broad peaks in the absorption spectra at nearly 750 and 490 nm [8]. A thorough UV-visible absorption measurements studies of the perovskite materials at various time intervals (in days) could indicate the change in the absorption of the band edge at 750 nm as well in the high energy transition around 480 nm. In the Fig. 2a presents the UV-visible absorption spectra for these 3D filmswhere these films are exposed to 52% relative humidity in dark at room temperature for several daysas shown in the inset of Fig 2a. [8] The intensity of the absorption bands (at 750 nm and 480 nm) decreases gradually indicating the consistent degradation of perovskites (Fig 2a). Furthermore, after

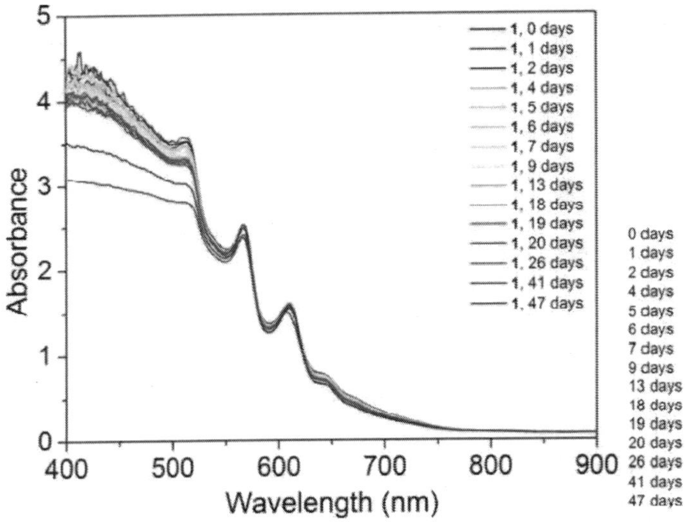

Fig 3. (a) UV-visible absorption spectra and (b) PXRD patters of of (PEA)2 (MA)2[Pb3I10] uponexposure to 52% relative humidity at various time intervals. [taken from ref. 8]

4 days of exposure to the 52% relative humidity, a sharp peak at 520 nm appears indicating the formation of the PbI_2. The instability of this $CH_3NH_3PbX_3$ materials is mainly due to the MA (methylammonium) group. This MA group is sensitive to moisture due to the hydrogen bond of the methylamine group and iodine atom. [12,16,17,18].

Powder X-ray diffraction (PXRD) also provides useful information related to state of solid materials. In the Fig. 2b the PXRD patterns of the $CH_3NH_3PbI_3$ films are shown immediately after preparation (0 day) as well as after various days after exposure to 52% relative humidity conditions. It is evident that, just after 4 days after the preparation of the films and exposure to 52% relative humidity conditions, a new phase emerges. Moreover, after 46 days of humidity exposure, this new phase dominates. This new phase could be easily correlated to PXRD patterns of the PbI_2 [16,17 ,8].As the efficiency of some of the 3D perovskite cells matches silicon solar cells with low-cost fabrication but the challenges are still unresolvable regarding the stability. The 3D perovskites are highly sensitive to moisture as well as high temperatures. By replacing the small organic group with bulky group i.e., replacing MA (methylammonium) group with BA (butyl ammonium) or PEA (phenyl ethyl ammonium) the stability increased (Table 1) [19,20].

2D PEROVSKITES

The 2D perovskite adopted the crystal structure of Ruddlesden-Popper type, general representationof this material $(RNH_2)_2A_{n\ 1}M_nX_{3n+1}$ (n = 1, 2, 3, 4, ∞), RNH_2 is bulky organic cation like BA or PEA, A is monovalent small organic ammonium cation, M is a divalent metal typically IV group element, X is halide. Number of layers depends on numerical n value. Different RNH_2, X groups havedifferent absorbance values as we saw for 3D materials [21,22,23,24]. The 3D perovskites are stableonly for two weeks at controlled environment while the 2D perovskites are stable at ambient conditions. This degradation of 2D materials can be monitored by the UV-visible absorption and PXRD measurements [8]. Fig 3a shows the UV-visible absorption measurements of 2D perovskites after exposing these films to 52% relative humidity. Clearly, no significant changes in the absorbanceof the multiple transitions even after exposure after 47 days. This confirms that, while the 3D perovskites are very sensitive to ambient atmosphere, their 2D counterparts are relatively less sensitive.

Table 1. Summary of different 3D & 2D perovskite-based solar cells.

Perovskite material	No. of layers	PCE %)	Device configuration	Stability	Ref
$(BA)_2(MA)_3Pb_4I_{13}$	4	12.52	ITO/PEDOT:PSS/ PVK/PCB M/Al	After 2250 h, 100% of PCE, dark, 65% RH	[25]
$(BA)_2(MA)_2Pb_3I_{10}$	3	6.89	ITO/PEDOT:PSS/ PVK/PCB M/BCP/Ag	100% PCE 24 days later, dark	[26]
$FAPbI_3\ _xBr_x$	-	17.9	PET/graphene/ PEDOT: PSSFAPbI3 xBrx/PCBM/Al	-	[27]
$(BA)_2(Cs_{0.05}MA_{0.95})_3Pb_4I_{13}$	4	13.68	FTO-e-TiO2/PVK/ spiro-OMeTAD/Au	After 1400 h, 89% of PCE, dark, 30% RH	[28]
$MAPbI_3$	-	10.06	FTO/TIO2/ MAPbI3/graphene		[29]
$(ThMA)_2(MA)_2PbI_{10}$	3	15.42	ITO/PEDOT:PSS/ PVK/PCB M/ BCP/Ag	90% PCE 1000 h later, dark	[30]
$(BA)_2(FA_{0.85}Cs_{0.15})_3(Pb_{0.6}\ Sn_{0.4})_4I_{13}$	4	9.3	ITO/PEDOT:PSS/ PVK/PCB M/BCP/Ag	After 2000 h, 47% of PCE	[31]

Perovskite material	No. of layers	PCE %)	Device configuration	Stability	Ref
MAPbI3	-	13.93	Quartz/graphene/ C60/MAPb I3/carbon	-	[32]
PDAMA3Pb4I13	4	13	ITO/PEDOT:PSS/ PVK/C60/ BCP/Ag	After 1000 h, 90% of PCE, dark	[33]
CA 2MA39Pb40I121	-	6.6	ITO/SnO2/PVK/ spiroOMeTAD/Au	After 264 h, 59% of PCE dark, 77%% RH	[34]
(PEA)MA4Pb5I 13	4	17.26	ITO/PTAA/PVK/ C60/BCP/A g	After 2000 h,96% of PCE, dark,	[35]
(BA,Gua)2MA4Pb5I16	5	15.86	ITO/PCP-Na/PVK/PCBM/ BCP/Ag	91% of PCE after 700h, dark, 55 RH	[36]
GAMA4Pb4I13	4	15.27	ITO/PEDOT:PSS/ PVK/PCB M/BCP/Ag	For 480 h, 75% of PCE, dark	[37]
(F-PEA)2(MA)4Pb 5I16	5	13.64	Device A with m-TiO2 on e-TiO2	After 76 h, 65%of PCE, dark, 70 C	[38]
(DA2PbI4)0.05MAPbI3	-	19.05	ITO/TiO2/PVK/ spiro-OMeTAD/ MoO3/Ag	After 60 days, 80% of PCE, dark	[39]
PEA15-SCN FASnI3	-	12.40	Glass/ITO/ PEDOT:PSS/PEA 15-SCNFASnI3/ ICBA/BCP/Ag	-	[40]
Ge doped FA0.98 EDA0.01 SnI3 (EA0.1)	-	13.24	Glass/FTO/ PEDOT:PSS/ Gedoped FA0.98EDA0.01SnI3 (EA 0.1)/C60/BCP/ Ag/Au	-	[41]
FA0.8 GA0.2 SnI3	-	10.24	-	-	[42]
FASnI3 (PAI)	-	12.11	Glass/ITO/PEDOT: PSS/A0.8GA0.2SnI3/ BAC2SnI4/C60/ BCP/Ag	After 20 days 137	[43]

Similar conclusions can be drawn from the PXRD pattern of $(PEA)_2(MA)_2[Pb_3I_{10}]$ stored at 52% relative humidity is measured for different time intervals as shown in Fig. 3b. It is clear that for 46 days the perovskite films does not undergo any degradation. While comparing to the 3D material under 52% relative humidity undergo degradation into PbI_2, very quickly.

CONCLUSIONS AND OUTLOOK

A detailed review of the stability and performance efficiency of 3D $CH_3NH_3PbI_3$ perovskites is presented here. The 3D perovskite films when exposed to relatively higher humidity conditions showed substantial degradation over time. This gradual degradation is measured by absorbance and PXRD studies. Further, we have also discussed on the improvement of stability of perovskite films when methylammonium cations in the 3D perovskite is replaced by a bulky cation i.e., phenylethylammonium cation and the structure changed from 3D $CH_3NH_3PbX_3$ to 2D $(PEA)_2(MA)_2[Pb_3I_{10}]$. In the 3D perovskites the methylamine is hydrophilic in nature and forms hydrogen bonds, so these films are sensitive to the moisture. While in the 2D perovskites, the bulky groups are hydrophobic and no hydrogen bonds are present, thus making them they are more stable towards humidity exposure. A detailed comparative analysis of 3D and 2D lead halide perovskites isgiven based on absorbance and PXRD data. It shows that 3D lead halides start decomposing to lead iodide after a few days, nonetheless, the 2D lead halide perovskites does not undergo any degradationfor around six weeks at 52% relative humidity exposure. Continued efforts are underway to consistently improve the stability and performance of the perovskite devices to match that of Si-basedsolar cells.

ACKNOWLEDGEMENTS

We thank the financial support from SERB, Govt. of India via sanction order SRG/2021/001358. We also thank VIT-AP University for other research facilities.

REFERENCES

[1] Y. Zhao, K. Zhu, Organic–inorganic hybrid lead halide perovskites for optoelectronic and electronic applications, Chem. Soc. Rev. 45 (2016) 655–689.

[2] J.Y. Kim, J.-W. Lee, H.S. Jung, H. Shin, N.-G. Park, High-Efficiency Perovskite Solar Cells, Chem. Rev. 120 (2020) 7867–7918.

[3] C.C. Stoumpos, C.D. Malliakas, M.G. Kanatzidis, Semiconducting Tin and Lead Iodide Perovskites with Organic Cations: Phase Transitions, High Mobilities, and Near-Infrared Photoluminescent Properties, Inorg. Chem. 52 (2013) 9019–9038.

[4] A. Kojima, K. Teshima, Y. Shirai, T. Miyasaka, Organometal Halide Perovskites as Visible-Light Sensitizers for Photovoltaic Cells, J. Am. Chem. Soc. 131 (2009) 6050–6051.

[5] M. Jeong, I.W. Choi, E.M. Go, Y. Cho, M. Kim, B. Lee, S. Jeong, Y. Jo, H.W. Choi, J. Lee, J.-H. Bae, S.K. Kwak, D.S. Kim, C. Yang, Stable perovskite solar cells with efficiency exceeding 24.8% and 0.3-V voltage loss, Science. 369 (2020) 1615–1620.

[6] G.-H. Kim, D.S. Kim, Development of perovskite solar cells with >25% conversion efficiency, Joule. 5 (2021) 1033–1035.

[7] N.H. Tiep, Z. Ku, H.J. Fan, Recent Advances in Improving the Stability of Perovskite Solar Cells, Adv. Energy Mater. 6 (2016) 1501420.

[8] I.C. Smith, E.T. Hoke, D. Solis-Ibarra, M.D. McGehee, H.I. Karunadasa, A Layered Hybrid Perovskite Solar-Cell Absorber with Enhanced Moisture Stability, Angew. Chem. Int. Ed. 53 (2014) 11232–11235.

[9] Y. Chen, Y. Sun, J. Peng, J. Tang, K. Zheng, Z. Liang, 2D Ruddlesden-Popper Perovskites for Optoelectronics, Adv. Mater. 30 (2018) 1703487.

[10] M. Shao, T. Bie, L. Yang, Y. Gao, X. Jin, F. He, N. Zheng, Y. Yu, X. Zhang, Over 21% Efficiency Stable 2D Perovskite Solar Cells, Advanced Materials. 34 (2022) 2107211.

[11] J.-H. Im, C.-R. Lee, J.-W. Lee, S.-W. Park, N.-G. Park, 6.5% efficient perovskite quantum-dot- sensitized solar cell, Nanoscale. 3 (2011) 4088.

[12] M.M. Lee, J. Teuscher, T. Miyasaka, T.N. Murakami, H.J. Snaith, Efficient Hybrid Solar Cells Based on Meso-Superstructured Organometal Halide Perovskites, Science. 338 (2012) 643–647.

[13] J.H. Heo, S.H. Im, J.H. Noh, T.N. Mandal, C.-S. Lim, J.A. Chang, Y.H. Lee, H. Kim, A. Sarkar, Md.K. Nazeeruddin, M. Grätzel, S.I. Seok, Efficient inorganic–organic hybrid heterojunction solar cells containing perovskite compound and polymeric hole conductors, Nature Photon. 7 (2013) 486– 491.

[14] H.-S. Kim, C.-R. Lee, J.-H. Im, K.-B. Lee, T. Moehl, A. Marchioro, S.-J. Moon, R. Humphry- Baker, J.-H. Yum, J.E. Moser, M. Grätzel, N.-G. Park, Lead Iodide Perovskite Sensitized All-Solid- State Submicron Thin Film Mesoscopic Solar Cell with Efficiency Exceeding 9%, Sci Rep. 2 (2012) 591.

[15] A. Mohammad Bagher, Types of Solar Cells and Application, AJOP. 3 (2015) 94.

[16] B. Brunetti, C. Cavallo, A. Ciccioli, G. Gigli, A. Latini, On the Thermal and Thermodynamic (In)Stability of Methylammonium Lead Halide Perovskites, Sci Rep. 6 (2016) 31896.

[17] G. Niu, X. Guo, L. Wang, Review of recent progress in chemical stability of perovskite solar cells, J. Mater. Chem. A. 3 (2015) 8970–8980.

[18] F. Zhang, D.H. Kim, K. Zhu, 3D/2D multidimensional perovskites: Balance of high performance and stability for perovskite solar cells, Current Opinion in Electrochemistry. 11 (2018) 105–113.

[19] C.C. Stoumpos, D.H. Cao, D.J. Clark, J. Young, J.M. Rondinelli, J.I. Jang, J.T. Hupp, M.G. Kanatzidis, Ruddlesden–Popper Hybrid Lead Iodide Perovskite 2D Homologous Semiconductors, Chem. Mater. 28 (2016) 2852–2867.

[20] Y. Chen, Y. Sun, J. Peng, W. Zhang, X. Su, K. Zheng, T. Pullerits, Z. Liang, Tailoring Organic Cation of 2D Air-Stable Organometal Halide Perovskites for Highly Efficient Planar Solar Cells, Adv. Energy Mater. 7 (2017) 1700162.

[21] A. Miyata, A. Mitioglu, P. Plochocka, O. Portugall, J.T.-W. Wang, S.D. Stranks, H.J. Snaith, R.J. Nicholas, Direct measurement of the exciton binding energy and effective masses for charge carriers in organic–inorganic tri-halide perovskites, Nature Phys. 11 (2015) 582–587.

[22] Y. Yang, D.P. Ostrowski, R.M. France, K. Zhu, J. van de Lagemaat, J.M. Luther, M.C. Beard, Observation of a hot-phonon bottleneck in lead-iodide perovskites, Nature Photon. 10 (2016) 53–59.

[23] Y. Yang, Y. Yan, M. Yang, S. Choi, K. Zhu, J.M. Luther, M.C. Beard, Low surface recombination velocity in solution-grown CH3NH3PbBr3 perovskite single crystal, Nat Commun. 6 (2015) 7961.

[24] Y. Yang, M. Yang, Z. Li, R. Crisp, K. Zhu, M.C. Beard, Comparison of Recombination Dynamics in CH 3 NH 3 PbBr 3 and CH 3 NH 3 PbI 3 Perovskite Films: Influence of Exciton Binding Energy, J. Phys. Chem. Lett. 6 (2015) 4688–4692.

[25] D.H. Cao, C.C. Stoumpos, O.K. Farha, J.T. Hupp, M.G. Kanatzidis, 2D Homologous Perovskites as Light-Absorbing Materials for Solar Cell Applications, J. Am. Chem. Soc. 137 (2015) 7843–7850.

[26] Y. Zheng, T. Niu, X. Ran, J. Qiu, B. Li, Y. Xia, Y. Chen, W. Huang, Unique characteristics of 2D Ruddlesden–Popper (2DRP) perovskite for future photovoltaic application, J. Mater. Chem. A. 7 (2019) 13860–13872.

[27] J.H. Heo, D.H. Shin, M.H. Jang, M.L. Lee, M.G. Kang, S.H. Im, Highly flexible, high-performance perovskite solar cells with adhesion promoted AuCl 3 -doped graphene electrodes, J. Mater. Chem. A. 5 (2017) 21146–21152.

[28] X. Zhang, X. Ren, B. Liu, R. Munir, X. Zhu, D. Yang, J. Li, Y. Liu, D.-M. Smilgies, R. Li, Z. Yang,T. Niu, X. Wang, A. Amassian, K. Zhao, S. (Frank) Liu, Stable high efficiency two-dimensional perovskite solar cells via cesium doping, Energy Environ. Sci. 10 (2017) 2095–2102.

[29] W. Wei, B. Hu, F. Jin, Z. Jing, Y. Li, A.A. García Blanco, D.J. Stacchiola, Y.H. Hu, Potassium- chemical synthesis of 3D graphene from CO 2 and its excellent performance in HTM-free perovskite solar cells, J. Mater. Chem. A. 5 (2017) 7749–7752.

[30] H. Lai, B. Kan, T. Liu, N. Zheng, Z. Xie, T. Zhou, X. Wan, X. Zhang, Y. Liu, Y. Chen, Two- Dimensional Ruddlesden–Popper Perovskite with Nanorod-like Morphology for Solar Cells with Efficiency Exceeding 15%, J. Am. Chem. Soc. 140 (2018) 11639–11646.

[31] D. Ramirez, K. Schutt, Z. Wang, A.J. Pearson, E. Ruggeri, H.J. Snaith, S.D. Stranks, F. Jaramillo, Layered Mixed Tin–Lead Hybrid Perovskite Solar Cells with High Stability, ACS Energy Lett. 3 (2018) 2246–2251.

[32] X. Meng, J. Zhou, J. Hou, X. Tao, S.H. Cheung, S.K. So, S. Yang, Versatility of Carbon Enables All Carbon Based Perovskite Solar Cells to Achieve High Efficiency and High Stability, Adv. Mater.30 (2018) 1706975.

[33] C. Ma, D. Shen, T.-W. Ng, M.-F. Lo, C.-S. Lee, 2D Perovskites with Short Interlayer Distance for High-Performance Solar Cell Application, Adv. Mater. 30 (2018) 1800710.

[34] R. Herckens, W.T.M. Van Gompel, W. Song, M.C. Gélvez-Rueda, A. Maufort, B. Ruttens, J. D'Haen, F.C. Grozema, T. Aernouts, L. Lutsen, D. Vanderzande, Multi-layered hybrid perovskites templated with carbazole derivatives: optical properties, enhanced moisture stability and solar cell characteristics, J. Mater. Chem. A. 6 (2018) 22899–22908.

[35] G. Wu, X. Li, J. Zhou, J. Zhang, X. Zhang, X. Leng, P. Wang, M. Chen, D. Zhang, K. Zhao, S. (Frank) Liu, H. Zhou, Y. Zhang, Fine Multi-Phase Alignments in 2D Perovskite Solar Cells with Efficiency over 17% via Slow Post-Annealing, Adv. Mater. 31 (2019) 1903889.

[36] X. Lian, J. Chen, Y. Zhang, S. Tian, M. Qin, J. Li, T.R. Andersen, G. Wu, X. Lu, H. Chen, Two- dimensional inverted planar perovskite solar cells with efficiency over 15% via solvent and interface engineering, J. Mater. Chem. A. 7 (2019) 18980–18986.

[37] Y. Zhang, J. Chen, X. Lian, M. Qin, J. Li, T.R. Andersen, X. Lu, G. Wu, H. Li, H. Chen, Highly Efficient Guanidinium-Based Quasi 2D Perovskite Solar Cells via a Two-Step Post-Treatment Process, Small Methods. 3 (2019) 1900375.

[38] F. Zhang, D.H. Kim, H. Lu, J.-S. Park, B.W. Larson, J. Hu, L. Gao, C. Xiao, O.G. Reid, X. Chen, Q. Zhao, P.F. Ndione, J.J. Berry, W. You, A. Walsh, M.C. Beard, K. Zhu, Enhanced Charge Transport in 2D Perovskites via Fluorination of Organic Cation, J. Am. Chem. Soc. 141 (2019) 5972–5979.

[39] X. Huang, Q. Cui, W. Bi, L. Li, P. Jia, Y. Hou, Y. Hu, Z. Lou, F. Teng, Two-dimensional additive diethylammonium iodide promoting crystal growth for efficient and stable perovskite solar cells, RSC Adv. 9 (2019) 7984–7991.

[40] X. Jiang, F. Wang, Q. Wei, H. Li, Y. Shang, W. Zhou, C. Wang, P. Cheng, Q. Chen, L. Chen, Z. Ning, Ultra-high open-circuit voltage of tin perovskite solar cells via an electron transporting layer design, Nat Commun. 11 (2020) 1245.

[41] K. Nishimura, M.A. Kamarudin, D. Hirotani, K. Hamada, Q. Shen, S. Iikubo, T. Minemoto, K. Yoshino, S. Hayase, Lead-free tin-halide perovskite solar cells with 13% efficiency, Nano Energy. 74 (2020) 104858.

[42] E. Jokar, P.-Y. Cheng, C.-Y. Lin, S. Narra, S. Shahbazi, E. Wei-Guang Diau, Enhanced Performance and Stability of 3D/2D Tin Perovskite Solar Cells Fabricated with a Sequential Solution Deposition, ACS Energy Lett. 6 (2021) 485–492.

[43] D. Cui, X. Liu, T. Wu, X. Lin, X. Luo, Y. Wu, H. Segawa, X. Yang, Y. Zhang, Y. Wang, L. Han, Making Room for Growing Oriented FASnI 3 with Large Grains via Cold Precursor Solution, Adv. Funct. Mater. 31 (2021) 2100931.

Chapter 6

Enhanced Magnetoelectric Coupling in Alternating Layers of Ferromagnetic and Ferroelectric Layers

V. K. Verma[1,*], V. R. Singh[2] and S. Jena[2]
Department of Physics, School of Advanced Sciences, VIT-AP University, Amaravati, 522237, A.P., India
[2]Department of Physics, Central University of South Bihar, Gaya, 824236, Bihar, India
*Corresponding Author: virendra.verma@vitap.ac.in**

INTRODUCTION

Multiferroics, a family of materials that show ferroelectric and ferromagnetic characteristics at the same time, have piqued the interest of researchers because of its potential uses in next-generation multifunctional devices [1-4]. If multiferroic materials have magnetoelectric (ME) coupling, an external magnetic field will produce electric polarization, whereas an external electric field will induce magnetization. A combination of these several ferroic orders may allow for close interaction among them, paving the path for cross-control of numerous ordered parameters. ME coupling is often weak in single-phase compound. Composite materials with alternating ferromagnetic and ferroelectric layers [5,6] or ferromagnetic nano-rods embedded in a ferroelectric matrix [5,7] may, on the other hand, display higher ME coupling. In a multiferroic material with co-existing ferroelectric and ferromagnetic ordering as well as strong enough ME coupling, the electric (magnetic) polarization should exhibit hysteresis as a function of the applied magnetic (electric) field. Such ME effects make multiferroic ME materials appealing for use in transducers, sensors, oscillators, filters, memory devices, phase shifters, and other applications. We can design a large number of combinations of ferroelectric and ferromagnetic materials to enhance the magnetoelectric coupling as shown in the Fig.1.

Here, we proposed an alternating layer of ferromagnetic and ferromagnetic layers to overcome the weak ME coupling effect. Strong magnetoelectric coupling between the two phases artificial ME composite will be developed on planner interface which will have the potential applications in novel multifunctional devices, such as sensors, memories, and transducers, etc. In this regards, we try to describe unsolved issues and new device applications which can be controlled both electrically and magnetically.

MATERIALS AND METHODS

The alternating layers of the ferromagnetic and ferromagnetic layers were prepared on Nb doped $SrTiO_3$ single crystal substrate by pulsed laser deposition (PLD) technique. The $BaTiO_3$ (BTO) and $NiFe_2O_4$ (NFO) multilayers were created using BTO and NFO ceramic targets. The thicknesses of the alternating layers of BTO and NFO were approximately 13.33 nm (9 layers), 24 nm (5 layers) and 40 nm (3 layers). The total thickness of the layers was 120 nm. The power density of the KrF laser was 1.5 J/cm^2 and the flowing oxygen pressure was 13 Pa. The quality of the films was verified by x-ray diffraction (XRD) and scanning electron microscopy (SEM). The beamline 11 at National Synchrotron Radiation Research Center (NSRRC) was used to characterize the x-ray absorption spectroscopy (XAS) and x-ray magnetic circular dichroism (XMCD) measurement. All experiments were carried out at room temperature using magnetic fields of up to 1 T. The base pressure of the XMCD chamber was maintained at 1 10^{9} Torr.

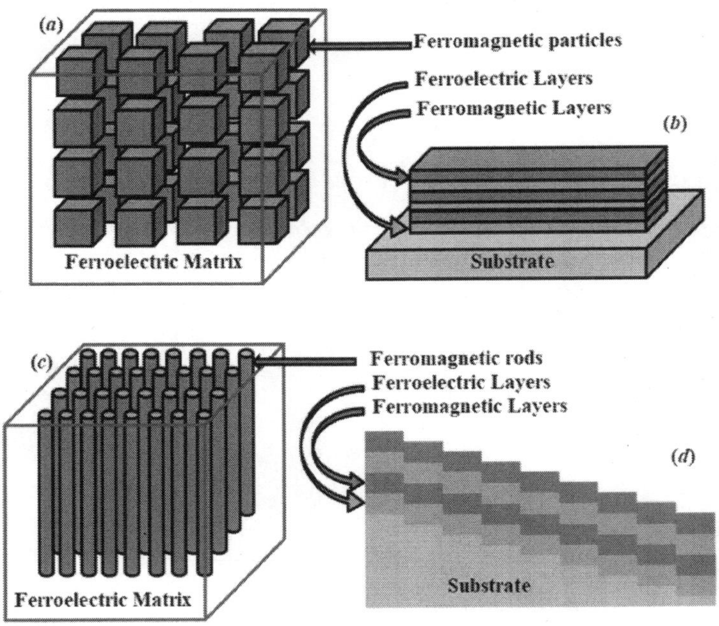

Figure 1. Schematic illustration of Ferromagnetic and ferroelectric materials (a) ferromagnetic particles in ferroelectric matrix (b) alternating ferromagnetic and ferroelectric layers, (c) ferromagnetic rods in ferroelectric matrix, and (d) stepped alternating ferromagnetic and ferroelectric layers

RESULTS AND DISCUSSION

The prepared samples of alternating layers of ferromagnetic (NFO) and ferroelectric (BTO) layers are demonstrated in Fig. 2. The total thicknesses of the all the layers in each samples were kept 120 nm. Scanning electron microscopy (SEM) was used for the measurement of thickness of layers [8]. First sample has 9 alternating layers (5 layers BTO and 4 layers NFO), second sample has 5 alternating layers (3 layers BTO and 2 layers NFO) and third sample has 3 alternating layers (2 layers BTO and 1 layer NFO). The total volumes of alternating layers of BTO and NFO in all the three samples are constant while the total interface areas in all the three samples are varying. The interface area to volume ratio is increasing with the increase of number of the layers. The Fe and Ni $L_{2,3}$-edge XAS and XMCD spectra of the alternating layers were carried out at room temperature. The absorption spectra of Fe and Ni are separated into L_3 ($2p_{3/2}$) and L_2 ($2p_{1/2}$) edges due to spin-orbit coupling. From the XAS and XMCD spectra of Fe and Ni, we found that the valence of Fe and Ni in the samples were 3+ and 2+, respectively [3]. The magnetic field dependence of the XMCD spectra of the samples shows residual magnetization at zero field which clearly indicates ferromagnetism at room temperature.

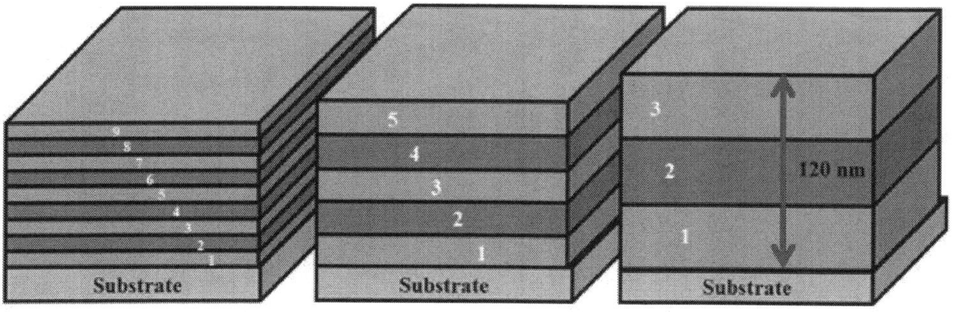

Figure 2. Alternating layers of ferromagnetic and ferroelectric layers on the substrate

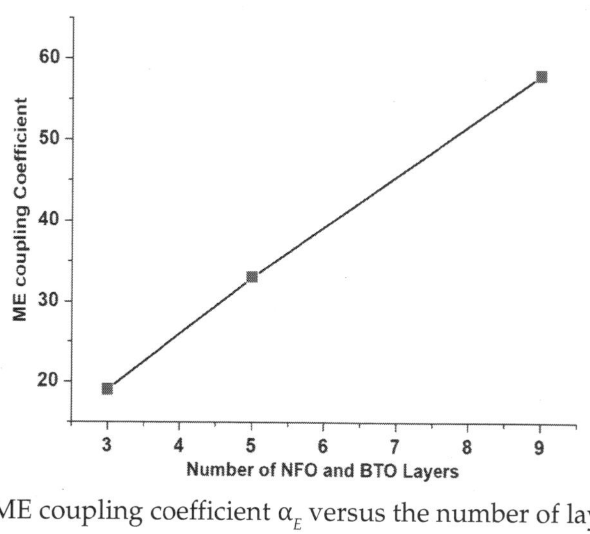

Figure 3. ME coupling coefficient α_E versus the number of layers

Figure 3 shows the ME coupling coefficient α_E versus the number of layers in the samples. Here, ME coupling coefficient α_E is defined by $\delta V/(\delta H t)$, where V is the induced voltage signal, δH is the magnetic excitation signal, and t is the total film thickness. The ME coupling coefficient α_E increases with the increase of number of layers. When the number of layers increases, the interface area to volume ratio increases and also the strain from BTO layer may impact the magnetization of the NFO.

Figure 4 shows the stepped surface of the $SrTiO_3$ (STO) substrates. This steeped surface on STO can be obtained by proper etching by the buffered oxide etch solution. On this stepped STO substrate, we can prepare the alternating layers of ferromagnetic and ferroelectric materials to get the enhanced ME coupling coefficient. The advantage of stepped surface is to increase the interface area and also the edge have different property like strain or bond break of Ti and O etc. which may affect the magnetization of ferromagnetic materials. In future, this work will motivate the many academic and industry researcher in related studies on multiferroic ME composites for micro-sensors such as $BaTiO_3/MnFe_2O_4$, $BaTiO_3/CoFe_2O_4$, and $BaTiO_3/MnCr_2O_4$ and deeper understanding of their underlying physics.

Figure 4. AFM image of stepped surface of SrTiO3

CONCLUSION

The alternating layers of ferromagnetic ($NiFe_2O_4$) and ferroelectric ($BaTiO_3$) samples were prepared by pulsed laser deposition technique. The Ni and Fe ions are in the 2+ and 3+ states in the samples. We found strong correlation between the ME coupling coefficient and number of ferromagnetic and ferroelectric layers. The ME coupling coefficient $_E$ increases with the increase of number of layers. The enhanced Me coupling coefficient is due to the increased interface area to volume ratio. The strain due to BTO at each interface may also affect the magnetization of the ferromagnetic layer. We also engineered the stepped interface on $SrTiO_3$ substrate. The alternating layers on the stepped substrate may enhance the ME coupling coefficient. Thus, we believe that this work can attract broad interest from research fields of magnetoelectric coupling between magnetostrictive and ferroelectric materials among academic and industry researcher.

ACKNOWLEDGMENT

We would like to thank K. Ishigami, G. Shibata, A. Fujimori, T. Harano, T. Kadono, F.-H. Chang, H.-J. Lin, D.-J. Huang and C. T. Chen for the technical support in the measurement of XAS and XMCD. We would also like to thank Yi Zhang, Jing Liu, Yuanhua Lin, Ce-Wen Nan for providing the samples. This work was supported by SERB-DST (ECR/2016/001741 & ECR/2017/000278).

REFERENCES

[1] W. Eerenstein, N. D. Mathur, and J. F. Scott. 2006. Multiferroic and magnetoelectric materials. Nature 442, 759-765.

[2] V. R. Singh, V. K. Verma, K. Ishigami, G. Shibata, Y. Yamazaki, A. Fujimori, Y. Takeda, T. Okane, Y. Saitoh, H. Yamagami, Y. Nakamura, M. Azuma, and Y. Shimakawa. 2013. Enhanced ferromagnetic moment in Co-doped BiFeO3 thin films studied by soft x-ray circular dichroism.J. Appl. Phys. 114, 103905.

[3] V. K. Verma, V. R. Singh, K. Ishigami, G. Shibata, T. Harano, T. Kadono, A. Fujimori, F.-H. Chang, H.-J. Lin, D.-J. Huang, C. T. Chen, Yi Zhang, Jing Liu, Yuanhua Lin, and Ce-Wen Nan. 2014. Origin of enhanced magnetoelectric coupling in $BaTiO_3/NiFe_2O_4$ multilayers studied by x-ray magnetic circular dichroism. Phys. Rev. B 89, 115128.

[4] Qian Song, Connor A. Occhialini, Emre Ergeçen, Batyr Ilyas, Danila Amoroso, Paolo Barone, Jesse Kapeghian, Kenji Watanabe, Takashi Taniguchi, Antia S. Botana, Silvia Picozzi, Nuh Gedik & Riccardo Comin. 2022. vidence for a single-layer van der Waals multiferroic. Nature 602, 601-605.

[5] H. Zheng, J.Wang, S. E. Lofland, Z. Ma, L.Mohaddes-Ardabili, T. Zhao, L. Salamanca-Riba, S. R. Shinde, S. B. Ogale, F. Bai, D. Viehland, Y. Jia, D. G. Schlom, M.Wuttig, A. Roytburd, and R. Ramesh. 2004. Multiferroic $BaTiO_3$-$CoFe_2O_4$ Nanostructures. Science 303, 661.

[6] J.Ma, J. Hu, Z. Li, and C-W. Nan. 2011. Recent Progress in Multiferroic Magnetoelectric Composites: from Bulk to Thin Films. Adv. Mater. 23, 1062.

[7] R. Ramesh and Nicola A. Spaldin. 2007. Multiferroics: progress and prospects in thin films. Nat. Mater. 6, 21.

[8] Jing Liu, Yi Zhang, Yuanhua Lin, and C. W. Nan 2009. Magnetoelectric coupling in $BaTiO_3/(NiFe_2O_4/BaTiO_3)_n$ (n = 1 , 2 , 3 , 4) multilayered thin films. J. Appl. Phys. 105, 083915.

Chapter 7

A Study on Silver Doped 2D MOS_2 Hybrid Nanostructure for Optoelectronic Device Applications

Christeena Thomas[1], Benazeera Beegum K A[1], Saranya Sasi[1], Asha A S[2], Reshmi R[1]
Union Christian College, Aluva, Kerala-683102
[2]Cochin University of Science and Technology (CUSAT), Ernakulam, Kerala-682022
E-mail : rreshmi@gmail.com

ABSTRACT

The characteristics evolution of materials with doping is an important area of research. In the current work, we have examined the changes in structural, morphological, optical, and electrical characteristics of hydrothermally synthesized 2D MoS_2 with 1% doping of $AgNO_3$ nanoparticles. The obtained XRD pattern is in good agreement with the crystal structure of MoS_2. The Raman analysis has shown characteristic peaks of 2H as well as 1T phases of MoS_2. Significant increases in surface area and layer separation with doping were observed from FESEM images of the materials. The EDAX spectrum has given substantiated evidence for the presence of molybdenum, sulfur, and silver particles in the synthesized material. Absorption studies using UV-Vis spectroscopy revealed enhancement in absorption and a slight increase in band gap due to the decrease in layer thickness. The I-V graph of the materials manifests a linear behavior with an effective increase in current for silver-decorated MoS_2. The temperature-dependent conductivity study demonstrated the semiconducting behavior of the materials. The improved conductivity in the silver-doped sample owing to the stable metallic nature of silver atoms.

INTRODUCTION

2D MoS_2 are the prime focus of research among various 2D nanomaterials for the last few years because of their novel optical, electrical, and mechanical properties.

The promising semiconducting nature and direct bandgap make MoS_2 an inevitable component of future high-end optoelectronic devices such as thin film transistors, photodetectors, solar cells, biosensors, etc [S Manzeli et al. (2017)]. All the transition metal dichalcogenides including MoS_2 can exist in different structural phases where the most common ones are the semiconducting 2H (trigonal prismatic) and metallic 1T (octahedral) phases. From recent studies, it is evident that the coexistence of 2H and 1T phases in a material can offer much stability and high carrier concentration [Y Yao et al. (2019)]. However, the pristine MoS_2 is active only at its edges and its basal planes are mostly inactive. Accordingly, for various applications meaningful modification of the material is necessary.

Doping with different atoms is an effective way to increase the properties of the material like exposed active sites and carrier mobility [A Nazneen et al. (2020)]. For 2D materials, doping can be done in different ways like Chemical Vapour Deposition (CVD), electrochemical method, thermal evaporation, hydrothermal method, etc [S A Han et al. (2015)]. Silver is a material with excellent optoelectronic properties and they are chemically stable and non-toxic. The effect of silver doping on the properties of 2D MoS_2 is less investigated and their electrical behavior is not studied in detail. The silver atom has an atomic radius of 144 pm which is less than the atomic radius of the molybdenum atom (154 pm) and can occupy both lattice sites and interstitial positions and can increase the overall performance of the material. Here in this study, the effect of 1% of $AgNO_3$ on the structural, morphological, optical, and electrical properties of 2D MoS_2 was investigated [A Nazneen et al. (2020), M Li et al (2019)].

MATERIALS AND METHODS

Chemicals

Sodium molybdate dihydrate ($MoNa_2O_4.2H_2O$), Thiourea (CH_4N_2S), and Silver Nitrate ($AgNO_3$). All the chemicals were reagent grade and were used as received.

Synthesis of pristine and Ag-doped MoS_2

Both MoS_2 and Ag-doped MoS_2 were synthesized using the hydrothermal synthesis technique. For the synthesis of MoS_2, 0.1M of $MoNa_2O_4.2H_2O$ and 0.1M of CH_4N_2S were stirred together in 70 ml of distilled water. Following 1 hour of stirring the mixture was transferred into a 100 ml Teflon beaker and heated at 200^0 C for 24 hours inside a sealed autoclave under high pressure. The derived precipitate was washed with distilled water and ethanol after cooling it to room temperature. The precipitate was dried at 200^0 C for 1 hour in a hot air oven. The Ag-doped MoS_2 was synthesized using the same method where 1% of $AgNO_3$ particles were incorporated with MoS_2 precursors during the stirring process as the silver source. For electrical studies, the material was coated on a PVDF membrane by Vaccum filtration. For that 50 mg of material was dispersed in 20ml distilled water and sonicated for 1 hour and Vaccum filtered using a PVDF membrane and dried at 50^0C for 1 hour. The obtained even film as such is used for conductivity measurements.

The XRD patterns were obtained using Bruker D8 Advance instrument, the Raman measurements were carried out using a Confocal Raman Microscope (WITec

Alpha 300RA) with a 532 nm DPSS laser, morphology of the samples and the EDAX measurements were studied using a Carl Zeiss Sigma field emission scanning electron microscope, UV/ Visible spectroscopy was done using SHIMADZU UV 1900i, and the electrical characteristics of the materials were investigated using Keithley 2450 Source Measuring Unit in resistive mode. The thickness of the material was obtained using DEKTAK 6M stylus profiler.

RESULTS AND DISCUSSION

The crystallite structure of the material was studied using the XRD spectrum (Fig 1). The distinct peaks of MoS_2 were observed corresponding to (002), (001), (103), (006), (105), (110), (108) planes respectively (JCPDS Data: 37-1492) for both pure and Ag-doped MoS_2.

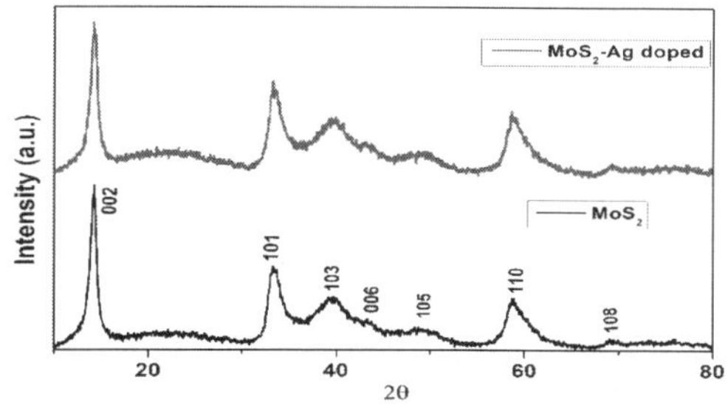

Figure 1. XRD spectrum of MoS_2 and Ag-doped MoS_2

There was no significant change in crystal structure observed for the Ag-doped sample implying a well-defined structure even after 1% doping. As the characteristic peaks of 1T and 2H phases of MoS_2 take the same position the coexistence of the phases cannot be defined using the XRD spectrum [Y Yao et al. (2019)].

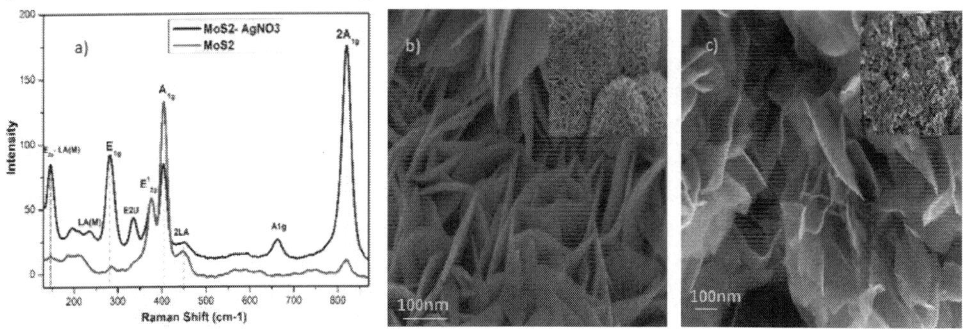

Figure 2. a) Raman spectra of MoS_2 and Ag-doped MoS_2 b) FESEM image of MoS_2 c) FESEM image of Ag-doped MoS_2.

Raman spectra (Fig 2 a) of the materials have shown characteristics peaks at 145 cm-1, 195 cm-1, 233 cm-1, 285 cm-1, 335 cm-1, 375 cm-1, 402 cm-1, 450 cm-1, 663 cm-1, and 819 cm-1. The peaks at 195 cm-1, 233 cm-1, and 335 cm-1 correspond to the 1T phase of MoS_2 and all the other peaks are attributed to the 2H phase of MoS_2. The signature first-order peaks of MoS_2 were observed at 375 cm-1 and 402 cm-1 representing the in-plane and out-of-plane vibration of molybdenum and sulfur atoms respectively [M Li et al (2019), B C Windom et al. (2011)]. The longitudinal acoustic (LA) modes represent the defects in the structure [A S Ahmed et al. (2018)]. The enhanced peak at 819 cm-1 is the combined effect of second order A_{1g} mode and the oxidation of the sample due to high laser intensity. The increase in the intensity of 1T peaks of the doped sample also proves the incorporation of metallic silver with MoS_2 [M Ikram (2020)].

The FFESEM images of MoS_2 (Fig 2 b)show a hierarchical structure where the sheets of materials are arranged like the petals in a flower. However, the entire symmetry of the material was disrupted by Ag doping (Fig 2 c) and has a more sheet-like structure with a noticeable decrease in thickness with a clear boundary.

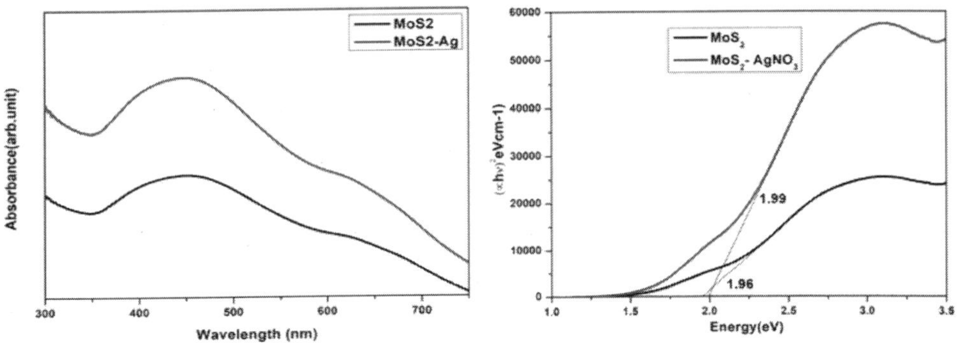

Figure 3: a) UV-Vis absorption spectrum of MoS_2 and Ag-doped MoS_2 b) Tauc plot of MoS_2 and Ag-doped MoS_2.

The UV-Vis absorption spectra (Fig.3) of MoS_2 and MoS_2- Ag samples show broad absorption spectra in the range of 350- 800nm with two absorption edges and a reasonable increase in absorption for the doped sample. The bandgap of the material was calculated using the Tauc equation (αE)1/n = C(E-Eg) Where C is a constant and α is the absorption coefficient. The value n is ½ for direct bandgap and 2 for indirect bandgap. The bandgap of MoS_2 was slightly increased from 1.96 eV to 1.99 eV upon silver doping indicating thickness dependent increase in band gap which is an inherent property of 2D materials, the observed bandgap is consistent with the direct bandgap value of MoS_2 [M Singh et al. (2018)].

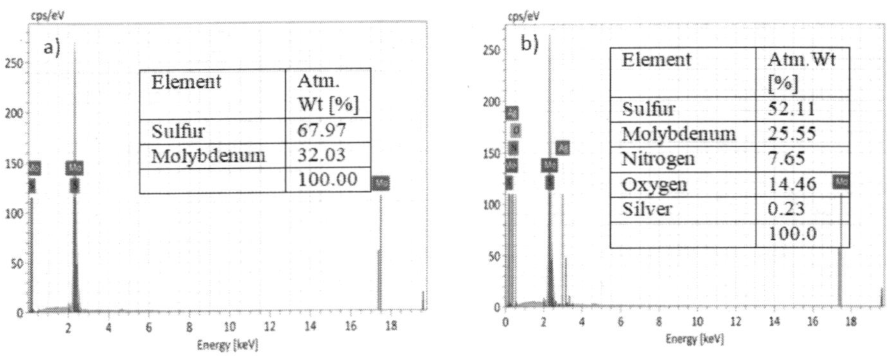

Figure 4: EDAX spectrum of a) MoS_2 b) Ag-doped MoS_2.

The EDAX spectrum (Fig 4) confirmed the presence of Mo, S, and Ag. The EDAX spectrum of the Ag-doped sample also shows the presence of oxygen and nitrogen which is a constituent of the silver source $AgNO_3$.

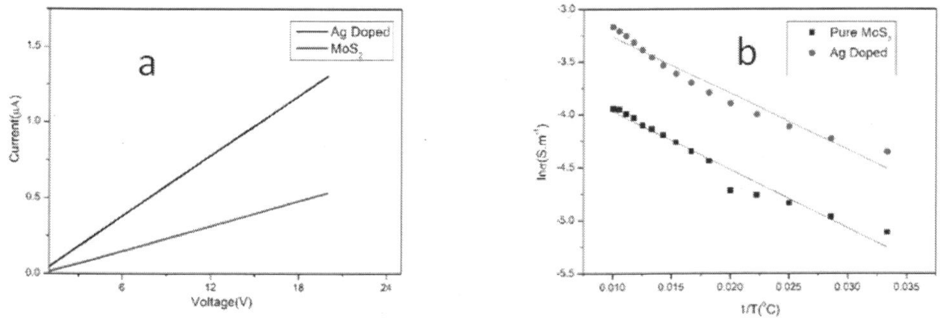

Figure 5: a) I-V characteristics of MoS_2 and Ag-doped MoS_2 b) Temperature-dependent conductivity of MoS_2 and Ag-doped MoS_2.

The I-V characteristics of the materials show linear nature with a greater increase in current for the Ag-doped sample. The temperature-dependent conductivity measurements demonstrated the semiconducting nature of the materials. A marked increase in conductivity for Ag-doped MoS_2 validates the substantial increase in carrier concentration due to the incorporation of silver atoms [S El-Mahalawy et al. (1977), T Thomas et al. (2019)]. The activation energy of the materials is calculated using the equation,

$$\sigma = \sigma_0 \exp(-\Delta E / K_B T)$$

Where is the conductivity, is the conductivity at 0^0 celsius, E is the activation energy, K_B is the Boltzman constant and T is the temperature.

Table 1: Data obtained from temperature-dependent studies

	Room temperature conductivity ($S.m^{-1}$)	Conductivity at 0^0C	Activation energy
MoS_2	0.006026	0.00096	0.00472
Ag Doped	0.012821	0.00215	0.0046

The Ag-doped MoS_2 showed a two-fold increase in conductivity and a decrease in activation energy with 1% doping. The conductivity of the material with various doping percent is ongoing.

ACKNOWLEDGMENT

We are grateful to DST SERB (Project No: CRG/2018/003785) for the financial support they have provided and to Cochin University of Science & Technology and SAIF, Mahatma Gandhi University for the characterization facilities.

CONCLUSION

The effect of silver doping on the structural, morphological, optical, and electrical properties of MoS_2 was examined in this study. The pure and doped samples were synthesized using the hydrothermal method. The structural analysis using XRD confirmed a well-defined crystal structure for both samples. The Raman analysis revealed the co-existence of 1T and 2H phases in both materials and a significant increase in metallic nature with doping. A well-structured surface with the least possible thickness is crucial for a device's performance. Herein morphological study a noticeable decrease in thickness and layer separation with clear boundary is manifested. The changes in bandgap value and a remarkable increase in absorption were shown in the UV-Vis spectrum. EDAX spectrum confirmed the presence of both dopant and base material. The linear dependence between current and voltage and the semiconducting nature of the materials with an effective increase of current were observed from electrical characterizations. Low power-consuming semiconducting devices with high carrier mobility and stability have paramount importance in future high-end optoelectronic device applications such as thin film transistors, photodetectors, solar cells, biosensors, etc. Here with 1% doping, the material showed enhancement in every property comprising the bifold increase in conductivity. Since the characteristics of Ag in $AgNO_3$ are very much dependent on its concentration a thorough investigation with different levels of doping and with different concentrations of $AgNO_3$ is necessary and that will be addressed in future studies. In conclusion, a close examination of the evolution of different 2D materials with doping is inevitable for future device-level applications.

REFERENCES

[1] Manzeli, S., Ovchinnikov, D., Pasquier, D., Yazyev, O.V. and Kis, A., 2017. 2D transition metal dichalcogenides. *Nature Reviews Materials*, 2(8), pp.1-15.

[2] Yao, Y., Ao, K., Lv, P. and Wei, Q., 2019. MoS2 coexisting in 1T and 2H phases synthesized by common hydrothermal method for hydrogen evolution reaction. *Nanomaterials, 9*(6), p.844.

[3] Nazneen, A., Khan, M.I., Naeem, M.A., Atif, M., Iqbal, M., Yaqub, N. and Farooq, W.A., 2020. Structural, morphological, optical, and photocatalytic properties of Ag-doped MoS2 nanoparticles. *Journal of Molecular Structure, 1220*, p.128735.

[4] Han, S.A., Bhatia, R. and Kim, S.W., 2015. Synthesis, properties and potential applications of two-dimensional transition metal dichalcogenides. *Nano Convergence*, 2(1), pp.1-14.

[5] Li, M., Cui, Z. and Li, E., 2019. Silver-modified MoS2 nanosheets as a high-efficiency visible-light photocatalyst for water splitting. *Ceramics International, 45*(11), pp.14449-14456.

[6] Windom, B.C., Sawyer, W.G. and Hahn, D.W., 2011. A Raman spectroscopic study of MoS2 and MoO3: applications to tribological systems. *Tribology Letters*, 42(3), pp.301-310.

[7] Ahmed, A.S., Xiang, W., Hu, X., Qi, C., Amiinu, I.S. and Zhao, X., 2018. Si3N4/ MoS2-PEDOT: PSS composite counter electrode for bifacial dye-sensitized solar cells. *Solar Energy, 173,* pp.1135-1143.

[8] Ikram, M., Khan, M.I., Raza, A., Imran, M., Ul-Hamid, A. and Ali, S., 2020. Outstanding performance of silver-decorated MoS2 nanopetals used as nanocatalyst for synthetic dye degradation. *Physica E: Low-dimensional Systems and Nanostructures, 124,* p.114246.

[9] Singh, M., Goyal, M., & Devlal, K. (2018). Size and shape effects on the band gap of semiconductor compound nanomaterials. *Journal of Taibah University for Science, 12*(4), 470-475.

[10] El-Mahalawy, S.H. and Evans, B.L., 1977. Temperature dependence of the electrical conductivity and hall coefficient in 2H-MoS2, MoSe2, WSe2, and MoTe2. *physica status solidi (b), 79*(2), pp.713-722.

[11] Thomas, T. and Mathew, A., 2019, July. Influence of post annealing on the properties of α-MoO3 nano rods grown on glass substrate by thermal evaporation. In *AIP Conference Proceedings* (Vol. 2115, No. 1, p. 030116). AIP Publishing LLC.

Chapter 8

Recent Advancements in The Photodegradation of Rhodamine B dyes

Keerthi Valsalan and Triveni Rajashekhar Mandlimath*

**Department of Chemistry, School of Advanced Scienes, VIT-AP University, Inavolu, Beside AP Secretariat, Amaravati, Andhra Pradesh, India – 522237*

Email: triveni.mandlimath@vitap.ac.in

INTRODUCTION

Synthetic dyes are the highly water-soluble complex organic molecules which exhibit wide range of colors. They are used in garments, ink, leather, beauty, pharmaceutical, food and beverage products. The largest segment of dye usage comes from textiles which owe approximately 72% of the global market in recent years. In textile industries, huge number of dyes is consumed for the dyeing process. After the dyeing process is over, 10-15% of the dissolved dyes are disposed into the water bodies such as rivers and sea. The presence of dyes in water causes water pollution. The extensive use and exposure to these dyes are carcinogenic, genotoxic and neurotoxic to humans. These dye molecules damage the physiological functions of aquatic animals and thus affect their lifetime. Therefore, the degradation of such pollutants is crucial for the sustainability of the ecosystem (Pereira and Alves (2012), Yusuf (2019),Kant (2012) and Gita(2017)).

Rhodamine B (RhB) is a xanthene dye that possess the chemical formula $C_{28}H_{31}ClN_2O_3$ and its structure is shown in figure 1. It gives bright red color in an aqueous solution (Sánchez-Valencia et al. (2010)). RhB is widely used in printing, dyeing in textile, paper and leather industries and also in bio imaging. The extensive use of RhB led to water pollution and cause health problems in humans such as skin irritation, eye irritation, nausea, vomiting and diarrhea. Since the dye is recalcitrant in nature, various physicochemical methods such as microfiltration, reverse osmosis, coagulation, adsorption and biological treatment have been implemented. But these processes are expensive, tedious, generate hazardous secondary pollutants and lead to incomplete removal of pollutants (Kestioglu et al.(2005)). To overcome such problems, photocatalytic degradation came into view. Photocatalytic degradation is an excellent method to remove non-biodegradable pollutants in water (Gusainet al. (2017)). In this chapter, we discuss various novel and modified photocatalysts utilized for the efficient degradation of RhB under UV as well as visible light irradiation.

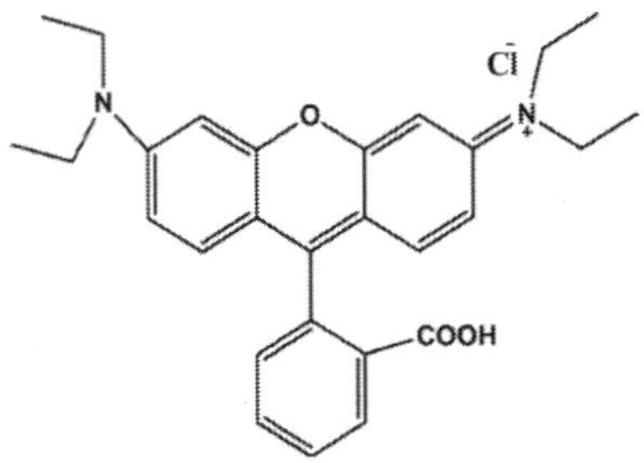

Figure 1. Structure of Rhodamine B

BASICS OF PHOTOCATALYSIS

Photocatalysis started in early 1911s where ZnO was reported for bleaching the Prussian Blue and photoreduction of Ag^+ to Ag. Later TiO_2, Nb_2O_5 were utilized for the photocatalytic reduction of Ag^+ to Ag. In 1972, Fujishima and Honda reported TiO_2 as a superior photocatalyst electrode to generate O_2 from water splitting by photoelectrochemical method under UV irradiation (Fujishima and Honda (1972)). Thus, the journey of various photocatalysts for different chemical reactions and photodegradation of various organic dyes started. TiO_2 photocatalyst is widely used because of its low cost and efficient photocatalytic activity. The wide band gap of TiO_2 (3.2 eV) makes it a UV-active photocatalyst. Since only 5% of solar light contains UV light, the utilization of TiO_2 asa photocatalyst is limited under visible light (Lee and Park (2013)).

In photocatalytic degradation, the light of energy hν (greater than the semiconductor band gap (E_g)) is irradiated onto the waste water containing the semiconductor photocatalyst. The photocatalyst absorbs photons followed by the generation of electrons and holes. The excited photoelectrons move toward the interface of the CB edge whereas, the holes migrate toward the interface near the valence band (VB). The oxidation and reduction reactions take place at the VB and CB interfaces respectively. During the redox reaction, reactive oxygen species (superoxide radicals and hydroxyl radicals) which are capable to degrade the pollutants are produced as shown in figure 2. The main advantage of photocatalysis is the production of harmless molecules like CO_2 and H_2O without any secondary pollutant formation (Zhu and Wang (2017)). For excellent photodegradation to occur, screening of reaction conditions such as light illumination intensity, pH, initial concentration of the dye solution and amount of photocatalyst are necessary.

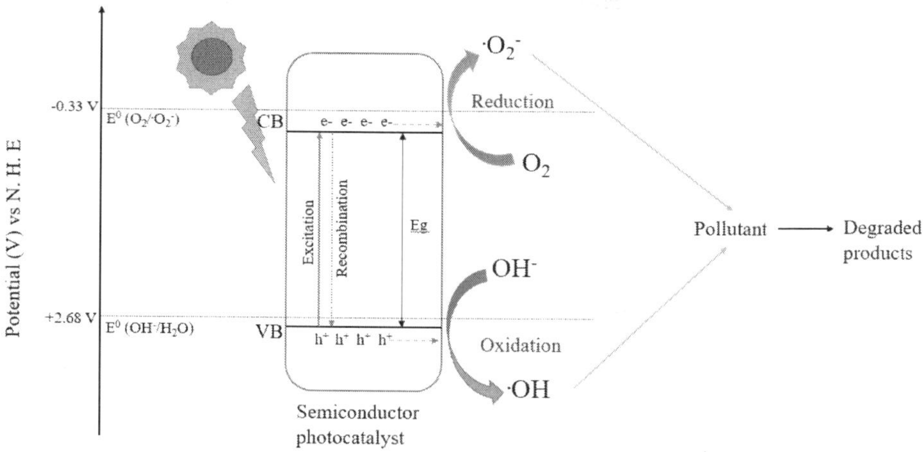

Figure 2. The photocatalytic mechanism for the degradation of pollutants

HOTOCATALYTIC DEGRADATION OF RHODAMINE B

Rhodamine B dye absorbs visible light and exhibits maximum absorbance at 553 nm. The photocatalytic degradation of RhB is indicated by the color fading of the solution and becomes colorless on complete removal. UV-Visible spectroscopy analysis of the dye solution at different time intervals reveals the photodegradation of the dye. The usual pattern of time-dependent UV-Visible spectra of RhB photodegradation in the presence of a photocatalyst is shown in figure 3.

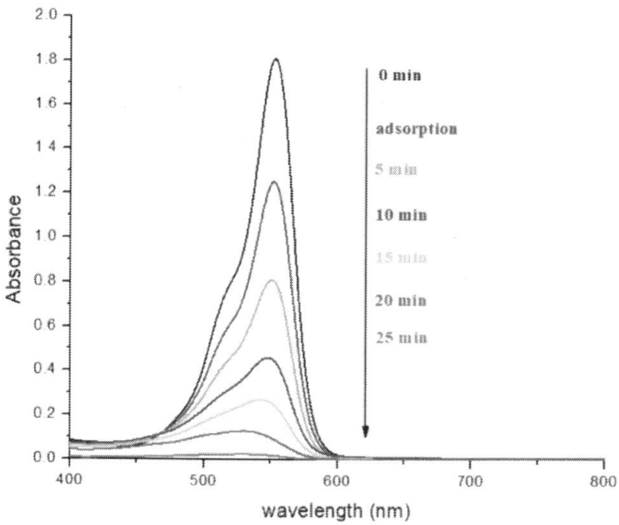

Figure 3. Time-dependent UV-Visible spectra of photocatalytic degradation of Rhodamine B dye

Several UV and visible light active semiconductor photocatalysts were reported by many researchers for the photodegradation of RhB. Photocatalysts namely, TiO_2, ZnO and $SrTiO_3$ exhibitedlow light harvesting ability, fast recombination of charge carriers and inefficient charge separation which limited them from efficient removal of RhB (Mills and Hunte (1997)). Therefore, many investigators have made an effort to develop low-cost, eco-friendly, and reusable photocatalysts with improved optical properties and photocatalytic activity. In this process, metal/ non-metal ion-doped photocatalysts and heterojunction photocatalysts have gained a lot of attention due to their enhanced photocatalytic efficiency for dye removal. The recently reported novel photocatalysts, doped and heterojunction photocatalysts which were studied for the RhB photodegradation are enlisted here and their details are given in table 1.

Hexagonal-shaped $TiSe_2$ nanoplates developed for the degradation of RhB by Ge et al. (2020), exhibited a narrow band gap 1.44 eV and low PL intensity. RhB underwent 96% degradation under visible light irradiation in 30 min in the presence of $TiSe_2$. The photocatalyst provided efficient photogeneration of electrons and holes, effective separation of the charge carriers and suppression of their recombination rate. $Bi_xSb_{2-x}S_3$ [x= 0.536, 1.09, 1.68], a ternary photocatalyst prepared by Dashairya et al. (2020), possessed nanorod morphology and band gaps ranges from 2.3 to 2.9 eV. The suppressed electron-hole recombination detoxified 95-99.5% RhB under visible light illumination in 30 min. The reusability and low-cost synthesis made this photocatalyst highly useful in degrading dye. Porous $Ti(MoO_4)_2$ nanospheres with 2.78 eV band gap showed 96% RhB degradation after 150 min under visible light irradiation (Mobeen et al. (2019)). BiF_3 proved to degrade 95.7% RhB in less than 1 h under solar light (Ritika et al.(2019)). ZnO with combined nanorods and nanoparticles morphology removed RhB almost completely under UV light (Nandi and Das (2019)).

Doped Photocatalysts

The electron-trapping nature of the dopant metal ions in semiconductor photocatalyst lead to enhance charge separation and suppress the recombination of electrons and holes. The formation of dopant energy level causes reduction in the band gap of the photocatalyst and thus enhances its photocatalytic efficiency in degrading RhB. Many researchers across the world have been reported various metal/non-metal ion-doped photocatalysts for RhB removal from water.

Qadir et al. (2022), reported Cr doped $La_2SrFe_2O_7$ prepared by sol-gel based Pechini method for RhB removal under visible light. $La_2SrFe_{1.8}Cr_{0.2}O_7$ showed a red shift in the light absorption, lower PL intensity and reduced band gap 1.66 eV. In the presence of H_2O_2 the catalyst exhibited 99% degradation in 30 min. La doping in WO_3 reduced the charge carrier recombination rate of the undoped WO_3. 4% La-doped WO_3 exhibited a reduction in the band gap from 2.83 eV (WO_3) to 2.46 eV. Under the illumination of visible light, 90% degradation of the dye took place in 90 min. The superoxide radicals produced from the photocatalyst were responsible for the degradation of RhB (Nehra et al. (2022)).Govindaraj et al. (2021) reported 3% Ce doped WO_3 showed excellent photocatalytic activity with 96% removal of RhB dye in 1 h. The band gap of 3% Ce-doped WO_3 was found to be 2.45 eV, showing a red shift in the light absorption peak and also low PL intensity. The Ce ions acted as the electron trapper in the reaction mechanism by which the photoelectrons and holes were able to generate the radicals efficiently for the photodegradation of RhB.

The reusability and reproducibility showed that the Ce-doped WO_3 can be used for pollutant removal applications. The doping of Ce into WO_3 gave better performance than La-doped photocatalyst.

The nanoflower-shaped MoS_2 doped with Sn with large surface area removed RhB completely in 20 min under visible light irradiation. The fast degradation efficiency of the catalysts is due to the porous nanoflower-like structure caused by Sn-doping and lyophilization (Nandigana et al. (2022)). 3% Zn doped CdS with 2.24 eV band gap could remove 93% RhB under sunlight. The photoelectrons of CdS transferred to Zn energy level due to the formation of Schottky junction. The enhanced photocatalytic activity was also because of the efficient charge separation and delay in their recombination. The photocatalyst showed stability under visible light for 3 cycles (Singh et al. (2022)). Zr-doped silver niobates possessed low PL intensity and a red shift in the UV-Visible absorption spectra. 10% Zr-doped $AgNbO_3$ with band gap 2.71 eV led to 77.5% degradation of RhB (Khor et al. (2022)).

Heterojunction Photocatalysts

The coupling of two or more photocatalysts with suitable electronic band structures leads to the formation of a heterojunction photocatalyst. This type of photocatalyst improves the light response range, electron/hole pair separation by various carrier transfer pathways and reduces the rate of recombination of the charge carriers. The photocatalytic activity for the degradation of RhB is enhanced by the use of heterojunction photocatalysts.

Many researchers across the globe have been reported verities of heterojunction photocatalysts for RhB removal. In this section, we highlight few of them. K. Saravanan et al., developed Ag/Fe_3O_4 nanocomposite functionalized with Jatropha curcas root at room temperature. The ferromagnetic Ag/Fe_3O_4 possessed band gap 2.8 eV removed 99% dye in 110 min under visible light irradiation. The reaction followed pseudo first-order kinetics with apparent rate constant value of 1.89 min^{-1}(Saravanan et al. (2022)). 20% Fe_3O_4/TiO_2 heterostructure was studied for RhB degradation by Madima et al. (2022). Compared to TiO_2, 20% Fe_3O_4/TiO_2 showed enhanced surface area, thermal stability, and extended absorption in visible light. The suppression of e^-/h^+ recombination in the formed heterojunction degraded 91% RhB in the presence of visible light. The photodegradation reaction fitted with pseudo first-order kinetics with rate constant of 0.0184 min^{-1}. The photocatalyst was found to be stable and reusable after 5 cycles (Ashiq et al. (2022)). The complete removal of RhB was carried out using g-C_3N_4/Ag@$CoWO_4$ composite with 2.3 eV band gap under sunlight in 2 hours. ZnO nanoparticles/ $PbCrO_4$ nanorods Type I heterojunction exhibited band gap 2.26 ± 0.01 eV (ZnO) and 3.21 ± 0.01 eV ($PbCrO_4$) and low PL intensity. The improved charge separation and reduced recombination rate of the charge carriers led to 95% RhB degradation under UV-A in 1 h and 80% under visible light after 3 h respectively (Hamza et al. (2022)).

Nanoplates of BiOCl-Graphene oxide heterojunction reported by Sarkar et al. (2022), exhibited visible light absorption ability and showed reduction in the band gap from 3.3 eV (BiOCl) to 3.1 eV (BiOCl-GO). The heterojunction formation enhanced the activity of BiOCl-GO and showed photodegradation efficiency of 99.59% within 60 min under visible light irradiation. Mesoporous BiOI nanoflowers with band gap 1.91 eV could degrade 63% RhB under visible light. The in-situ partial conversion of the catalyst BiOI into BiOCl on addition of HCl increased the photodegradation of RhB to 99.2%. The addition of Cl- ions in RhB solution thus helped to upgrade the photocatalytic efficiency of BiOI (Chakraborty et al. (2021)).

The CTAB and SDS capped BiOCl possessed band gap 3.47 eV and 3.61 eV respectively. RhB was removed in 40 min successfully in the presence of these catalysts (Narenuchet al. (2021)). PVP-capped Gd-doped ZnS showed slight red shift in the light absorption capacity and reduced the band gap of ZnS. The 1% Gd-doped photocatalyst degraded 90% of RhB in 10 min under both UV and visible light irradiation. Although the photocatalyst shows high activity but undergoes photo corrosion (Dhir (2021)).

Enhancement in the photocatalytic activity of WO_3 by the formation of heterojunction with AgBr was studied by Piriyanon et al. (2021) The heterojunction with 2.93 eV band gap experienced reduced recombination of charge carriers and increased separation of electrons/holes. WO_3/AgBr showed complete degradation of RhB after 240 min under UV light and after 180 min under solar light irradiation. 95% degradation of RhB was achieved by 0.25 wt% Pt/TiO_2 nanocomposite under UV and sunlight irradiation. The optical band gap of the photocatalyst was 3.06 eV which is lower than the band gap of TiO_2. A Schottky barrier was created between TiO_2 and Pt and therefore the photogenerated electrons transferred to Pt energy level and efficient generation of superoxide radicals took place resulting in improved photocatalytic activity of TiO_2(Abdel-Khaleket al. (2022)). Stable and recyclable poly [ANE+NPMI]-TiO_2 microspheres exhibited 340.6 m^2/g BET surface area and performed almost complete degradation of RhB under UV light illumination within 90 min (Li et al. (2021)).

Tran et al. (2021) reported $Cu_{0.5}Mg_{0.5}Fe_2O_4$ spinel doped with TiO_2. The catalyst with 2.86 eV bang gap lowered the rate of charge carrier recombination and exhibited 100% removal of RhB dye after 180 min in the presence of visible light. The photocatalytic mechanism studies revealed that the photogenerated electrons transferred from the CB of spinel to the CB of TiO_2 and also photogenerated holes transferred from the VB of TiO_2 to the VB of the spinel. Successive generation of superoxide radicals and hydroxyl radicals completely degraded RhB dye. Thus, the hybrid material is as an efficient photocatalyst for the removal of organic dye [. Chen et al. (2020) developed a heterojunction photocatalyst by growing TiO_2 on CNT surface by CTAB. The thermal stability of TiO_2 was improved by the introduction of CNT and obtained 2.28 eV band gap. The formation of heterojunction lowered the recombination of charge carriers and improved the photoelectrons migration from the CB of TiO_2 to CNT. 10% CNT-TiO_2 heterojunction proved 100% removal of the dye in 25 min under UV light irradiation.Although the heterojunction exhibited high visible light absorption capacity because of CNT, the composite showed low performance in the degradation reaction under visible light.

Table 1: List of recently used photocatalysts for RhB degradation

Photocatalyst (mg)	Rhodamine B dye	Removal efficiency	The apparent rate	Reusability of the photocatalyst	Reference
Ag/Fe_3O_4 (10 mg)	50 mL of 10 ppm	99% in 110 min	1.89	5 cycles	Saravanan et al.(2022)
Fe_3O_4/TiO_2 (50 mg)	100 mL of 10 ppm	91% after 120 min	0.0184	5 cycles	Madima et al. (2022)
g-C_3N_4/Ag@ $CoWO_4$	100 mL of 10 ppm, 15 mM H_2O_2, pH=6	100% in 120 min	0.03	5 cycles	Ashiq et al. (2022)
$La_2SrFe_{1.8}Cr_{0.2}O_7$ (62.5 mg)	50 mL, pH=3, 0.5 mL H_2O_2 (27 wt%)	99% in 30 min	-	-	Qadir et al. (2022)
$W_{1-x}La_xO_3$ (10 mg)	150 mL of 10 ppm, pH=6.5	90% in 90 min	0.0204	-	Nehra et al. (2022)
Lyophilized Sn-doped MoS_2 (50 mg)	50 mL of 10 ppm	100% in 20 min	0.2120	4 cycles	Nandigana et al. (2022)
Zn-doped CdS (20 mg)	100 mL of 0.5 mg/100 mL	85% under UV, 93% under visible light	0.0081	3 cycles	Singh et al. (2022)
Zr-doped $AgNbO_3$ (20 mg)	50 mL of 10 ppm	77.5% in 5 h under visible light	-	-	Khor et al. (2022)
$ZnO/PbCrO_4$ (100 mg)	100 mL of 10^{-5} M	95% in 1 h under UV-A, 80% after 3 h under visible light	-	-	Hamza et al. (2022)
BiOCl-GO (30 mg)	40 mL of 10^{-5}M	99.59% in 60 min	0.1333	3 cycles	Sarkar et al. (2022)
BiOI/BiOCl (50 mg)	5 ppm, pH=3	99.2% in 1 h	0.1651	4 cycles	Chakraborty et al. (2021)
Ce-doped WO_3 (5 mg)	50 mL of 20 mg/L	96% in 30 min	0.0318	3 cycles	Govindaraj et al. (2021)

Photocatalyst (mg)	Rhodamine B dye	Removal efficiency	The apparent rate	Reusability of the photocatalyst	Reference
CTAB or SDS capped BiOCl (50 mg)	200 mL of 10 ppm	98%, 100% under solar light	0.0965, 0.0051	3 cycles	Narenuch et al. (2021)
PVP capped Gd-doped ZnS (100 mg)	100mL of 25 ppm	90% under UV and Visible light	0.0173	3 cycles	Dhir (2021)
WO_3/AgBr (50 mg)	200 mL of 10 ppm	100% under both UV and solar light, 86% under visible light	0.0078 (UV), 0.0256 (solar light), 0.0123 (visible)	5 cycles	Piriyanon et al. (2021)
ZnO/Bi_2MoO_6 (50 mg)	200 mL of 10 ppm	100% in 240 min	0.0549	5 cycles	Chankhanittha and Nanan (2021)
Poly [ANE+NPMI]-TiO_2 (20 mg)	20 mL of 50 ppm	95% in 120 min	0.0284	7 cycles	Li et al. (2021)
$Cu_{0.5}Mg_{0.5}Fe_2O_4$-TiO_2 (20 mg)	20 mL of 10 ppm	100%	0.0139	5 cycles	Van Tran et al. (2021)
$NiFe_2O_4$@Sn-HAP (50 mg)	50 mL of 10 ppm, 5mL H_2O_2 (10%)	99.2%	0.0877	-	Das et al. (2021)
$TiSe_2$ (30 mg)	30 mL of 10 ppm	96% in 90 min	0.0546	-	Ge et al. (2020)
SnS_2-CdO (6 mg)	100 mL of 0.025 M	86.11%	-	-	Srivind et al. (2020)
TiO_2-CNT (10 mg)	100 mL of 10 ppm	100% in 25 min	-	4 cycles	Chen et al. (2020)
$Bi_xSb_{2-x}S_3$ (5 mg)	10 ppm	95-99.5%	-	5 cycles	Dashairya et al. (2020)
$Ti(MoO_4)_2$ (75 mg)	100 mL of 10 ppm	96% after 150 min	-	-	Mobeen et al. (2019)
Pt/TiO_2 (0.1g)	100 mL of 3×10^{-5} mol/L	95%	-	5 cycles	Abdel-Khalek et al. (2022)

Agglomerated quasi-spherical $NiFe_2O_4$ nanoparticles encapsulated by Sn^{2+} doped HAP rods possessed large surface area with 1.48 eV band gap. Under visible light illumination, 99.2% and 88.4% removal of dye was achieved in 40 min with and without H_2O_2 respectively (Das et al. (2021). The heterojunction of ZnO nanoparticles with flower-shaped Bi_2MoO_6 was fabricated by Chankhanittha and Nanan (2021) which exhibited low PL intensity. RhB was degraded to 100% after 240 min under visible light illumination by the heterojunction photocatalyst. The formation of Type II heterojunction enhanced the photogenerated electrons and holes separation, migration and reduced electron-hole recombination rate resulted in high photocatalytic efficiency.

SnS_2-CdO heterojunction with 2.4 eV band gap possessed sulphur or oxygen vacancies. 86.11% degradation of RhB was achieved after 240 min in the presence of the heterojunction. The photogenerated electrons from the CB of SnS_2 transferred to the CB of CdO and thus, superoxide radicals were generated at the interface that degraded the dye. Simultaneously, the photogenerated holes in VB of both phases produced hydroxyl radicals which also degraded the dye (Srivind et al. (2020)). A heterojunction of Sm_2O_3-TiO_2 with large surface area and 2.6 eV band gap showed 94% removal of RhB under visible light illumination (Kaviyarasu et al.(2020)).

CONCLUSION

In recent decades, the development of highly efficient photocatalysts to remove dyes is one of the most active research areas. In this review, metal/non-metal doped and heterojunction photocatalysts have gained more attention due to their large surface area, low-band gap, improved light absorption capacity, increased photogenerated electrons/holes separation and reduced recombination rate. Magnetic photocatalysts are found to be easily separated using external magnets. Stable and reusable photocatalysts are promising materials for dye removal. Most of the photocatalysts effectively remove the dye in less time. This book chapter provides insight into the Rhodamine B dye degradation using varieties of photocatalysts under optimal reaction conditions. This will help the readers to understand the existing gaps for the development of novel photocatalysts for the complete removal of the dye.

REFERENCES

[1]Abdel-Khalek. M. H, Ahmed. M. A, Abdel-Messih. M. F and El-Shahat. F. 2022. Synthesis of mesoporous Pt/TiO_2 nanoparticles by incipient wetness route for photocatalytic degradation of rhodamine B and methyl orange dyes under UV and sunlight radiations, Mater. Sci. Technol., 5: 334-343.

[2] Ashiq. H et al. 2022. G-C_3N_4/Ag@$CoWO_4$: A novel sunlight active ternary nanocomposite for potential photocatalytic degradation of rhodamine B dye, J. Phys. Chem. Solids, 161: 110437.

[3] Chakraborty. M et al. 2021. Synthesis of mesoporous BiOI flower and facile in-situ preparation of BiOI/BiOCl mixture for enhanced photocatalytic degradation of toxic dye, Rhodamine-B, J. Photochem. Photobiol., 8: 100077.

[4] Chankhanittha. T and S. Nanan. (2021). Visible-light-driven photocatalytic degradation of ofloxacin (OFL) antibiotic and Rhodamine B (RhB) dye by solvothermally grown ZnO/Bi_2MoO_6 heterojunction, J. Colloid Interface Sci., 582: 412-427.

[5] Chen. Y., J. Qian, N. Wang, J. Xing and L. Liu. 2020. In-situ synthesis of CNT/TiO_2 heterojunction nanocomposite and its efficient photocatalytic degradation of Rhodamine B dye, Inorg. Chem. Commun., 119: 108071.

[6] Das. K. C., S. S. Dhar, D. G. Thakurata and J. Das. 2021. Sn (II) inserted on hydroxyapatite encapsulated nickel ferrite ($NiFe_2O_4$@HAp-Sn^{2+}): A novel nanocomposite for the effective photo-degradation of rhodamine B dye, J. Clean. Prod., 290: 125172.

[7] Dashairya. L., A. Mehta, P. Saha and S. Basu. 2020. Visible-light-induced enhanced photocatalytic degradation of Rhodamine-B dye using $Bi_xSb_{2-x}S_3$ solid-solution photocatalysts, J. Colloid Interface Sci., 561: 71-82.

[8] Dhir. R. 2021. Synthesis, characterization and applications of Gadolinium doped ZnS nanoparticles as photocatalysts for the degradation of dyes (Malachite Green and Rhodamine B) and as antioxidants, Chem. Phys., 3: 100027.

[9] Fujishima. A and K. Honda. 1972. Electrochemical photolysis of water at a semiconductor electrode, Nature, 238: 37-38.

[10] Gita. S., A. Hussan and T. G. Choudhury. 2017. Impact of textile dyes waste on aquatic environments and its treatment, Environ. Ecol., 35 (3C): 2349–2353.

[11] Ge. W et al. 2020. Trioctylphosphine (TOP)-free synthesis of $TiSe_2$ plates for enhanced photocatalytic degradation performance of Rhodamine B dyes, Solid State Sci., 103: 106189.

[12] Govindaraj. T et al. 2021. Enhanced visible-light-driven photocatalytic activity of Ce doped WO_3 nanorods for Rhodamine B dye degradation, Mater. Lett., 305: 130705.

[13] Gusain. R., K. Gupta, P. Joshi and O. P. Khatri. 2019. Adsorptive removal and photocatalytic degradation of organic pollutants using metal oxides and their composites: A comprehensive review, Adv. Colloid Interface Sci., 272: 102009.

[14] Hamza. M.A., S. A. Abd El-Rahman and Z. M. Abou-Gamra. 2022. Facile one-pot solid-state fabrication of a novel binary nanocomposite of commercial ZnO and commercial $PbCrO_4$ with enhanced photocatalytic degradation of Rhodamine B dye, Opt. Mater., 124: 111987.

[15] Kaviyarasu. K et al. 2020. High performance of pyrochlore like $Sm_2Ti_2O_7$ heterojunction photocatalyst for efficient degradation of rhodamine-B dye with wastewater under visible light irradiation, J. King Saud Univ. Sci., 32: 1516-1522.

[15] Kant. R. 2012. Textile dyeing industry an environmental hazard, Nat. Sci., 4: 22-26.

[16] Kestioglu. K., T. Yonar and N. Azbar. 2005. Feasibility of physico-chemical treatment and advanced oxidation processes (AOPs) as a means of pretreatment of olive mill effluent (OME), Proc. Biochem., 40: 2409–2416.

[17] Khor. C et al. 2022. Zr-doped silver niobates for photocatalytic degradation of methylene blue and rhodamine B dyes, Optik, 267: 169732.

[18] Lee S. Y and S. J. Park. 2013. TiO_2 photocatalyst for water treatment applications, J. Ind. Eng. Chem., 19: 1761-1769.

[19] Li. X., S. Raza and C. Liu. 2021. Preparation of titanium dioxide modified biomass polymer microspheres for photocatalytic degradation of rhodamine-B dye and tetracycline, J. Taiwan Inst. Chem. Eng., 122: 157-167.

[20] Madima. N et al. 2022. Fabrication of magnetic recoverable Fe_3O_4/TiO_2 heterostructure for photocatalytic degradation of rhodamine B dye, Inorg. Chem. Commun., 145: 109966.

[21] Mills. A and S. L. Hunte. 1997. An overview of semiconductor photocatalysts, J. Photochem. Photobiol. A. Chem., 108: 1-35.

[22] Mobeen A. et al. 2019. Investigation on antibacterial and photocatalytic degradation of Rhodamine-B dye under visible light irradiation by titanium molybdate nanoparticles prepared via microwave method, Surf. Interfaces, 17: 100381.

[23] Nandi. P and D. Das. 2019. Photocatalytic degradation of Rhodamine-B dye by stable ZnO nanostructures with different calcination temperature induced defects, App. Surf. Sci., 465: 546-556.

[23] Nandigana. P et al. 2022. Lyophilized tin-doped MoS_2 as an efficient photocatalyst for overall degradation of Rhodamine B dye, J. Alloys Compd., 907: 164470.

[25] Narenuch. T., T. Senasu, T. Chankhanittha and S. Nanan. 2021. Solvothermal synthesis of CTAB capped and SDS capped BiOCl photocatalysts for degradation of rhodamine B (RhB) dye and fluoroquinolone antibiotics, J. Solid State Chem., 294: 121824.

[26] Nehra. P., A. Khatri, A. Kumar, S. Singh and P. S. Rana. 2022. Synthesis and characterization of rare earth metal doped tungsten trioxide photocatalyst for degradation of Rhodamine B dye. J. Rare Earths. doi.org/10.1016/j.jre.2022.07.007.

[27] Piriyanon. J et al. 2021. Performance of sunlight responsive WO_3/AgBr heterojunction photocatalyst toward degradation of Rhodamine B dye and ofloxacin antibiotic, Opt. Mater., 121: 111573.

[28] Pereira. L. and M. Alves. 2012. Dyes-environmental impact and remediation, Environ. Prot. Strateg. Sustain. Dev. 111–162.

[29] Qadir I. et al. 2022. Singh, Influence of Cr substitution on structural, optical, and magnetic properties of $La_2SrFe_2O_7$: Their application as a photocatalyst for the degradation of Rhodamine B dye, Ceram. Int., 48: 33267-33279.

[30] Ritika et al. 2019. BiF_3 octahedrons: A potential natural solar light active photocatalyst for the degradation of Rhodamine B dye in aqueous phase, Mater. Res. Bull., 112: 376-383.

[31] Sánchez-Valencia J. R. et al., 2010. Excitation transfer mechanism along the visible to the near-IR in rhodamine J-heteroaggregates, Chem. Commun., 46: 4372-4374.

[32] Saravanan. K et al. 2022. Green synthesis and characterization of Ag and Ag/Fe_3O_4 nanocomposites for antimicrobial effect and rhodamine- B dye degradation, J. Indian Chem. Soc., 99: 100575.

[33] Sarkar. R et al. 2022. Hydrothermal synthesis of GO wrapped BiOCl nanosheet and its application in visible light assited catalytic degradation of Rhodamine B dye, Mater. Chem. Phys., 279: 125796.

[34] Singh. A., D. Singh, B. Ahmed and A. K. Ojha. 2022. Sun/UV-light driven photocatalytic degradation of rhodamine B dye by Zn doped CdS nanostructures as photocatalyst, Mater. Chem. Phys., 277: 125531.

[35] Srivind. J et al. 2020. Visible light irradiated photocatalytic performance of SnS_2-CdO nanocomposite against the degradation of rhodamine B (cationic) and congo red (anionic) dyes, Mater. Sci. Eng. B, 255: 114530.

[36] Van Tran. C et al. 2021. New TiO_2-doped Cu–Mg spinel-ferrite-based photocatalyst for degrading highly toxic rhodamine B dye in wastewater, J. Hazard. Mater., 420: 126636.

[37] Yusuf. M. 2019. Synthetic dyes: A threat to the environment and water ecosystem, Text. Cloth. Environ. Concerns Solut., 11–26.

[38W] Zhu. S and D. Wang. 2017. Photocatalysis: basic principles, diverse forms of implementations and emerging scientific opportunities, Adv. Energy Mater., 7: 1700841.

Chapter 9

An Insight into The Synthetic Methods of Carbon Dots and Hetero Atom Doped Carbon Dots

Reshmi M K Nair and Triveni Rajashekhar Mandlimath*

Department of Chemistry, School of Advanced Sciences, VIT-AP University, Inavolu, Beside AP Secretariat, Amaravati, Andhra Pradesh, India-522237

Email: triveni.mandlimath@vitap.ac.in

INTRODUCTION

Carbon dots are the most important ones in the world of zero-dimensional carbon nano materials. They have a spherical or quasi spherical structure having size in between 1-10 nm. The core of carbon dots (C-dots) consists of a mixture of sp^2 and sp^3 hybrid carbon atoms and their surface contain many functional groups such as hydroxyl, amino and carboxyl. C-dots were accidentally spotted by Xu et al., during the purification of single walled carbon nanotubes in 2004 (Xu et al., (2004)). Immediately after two years, for the first time Sun et al., developed the first-generation C-dots (Sun et al., (2006)). Since then, C-dots have been emerged as a dominant zero-dimensional material in the field of research due to their vast applications. These little ones are the building blocks for future devices. While comparing with the traditional nanomaterials, they have the upper hand due to their unique properties such as biocompatibility, optical property, easy surface functionalization, hydrophilic nature and less cytotoxicity. Being inexpensive, C-dots are applicable in catalysis, optoelectronics, energy conversions, sensing of heavy metal ions, bioimaging of cancer cells, drug delivery and waste water treatment. Even though C-dots exhibit excellent bio-applications, due to their poor quantum yield and interference by other fluorescent background materials, their applications are limited. Most reported C-dots show blue fluorescence which is not suitable for biological studies (Radnia et al., (2020), Zhang and Yu, (2016)).

Hetero atom doped C-dots aim to overcome both these problems. Hetero atoms can be Nitrogen, Sulphur, Boron and Silicon. Doping of C-dots with hetero atom enhance their quantum yield and electronic properties by the introduction of new energy levels. These energy levels are introduced by overlapping the atomic orbitals of carbon atom with the atomic orbitals of hetero atom and gives emission colour other than blue (Wang et al., (2022)). The quantum yield and optoelectronic properties of C-dots and hetero atom doped C-dots can also vary with the synthetic methods, precursors used and reaction conditions. The exact reason for this variation and proper mechanism of these properties is still in debate. C-dots are synthesized from both inorganic (Hettiarachchi et al., (2019)) and organic (Mewada et al., (2014)) sources using different synthetic methods. C-dots prepared from naturally available materials such as fruit juice (Mehta et al., (2015), Huang et al., (2013)), fruit peels (Tyagi et al., (2016), Surendran et al., (2021)), vegetables (Das et al., (2019)), leaves (Maruthupandi et al., (2022), Kalanidhi and Nagaraaj, (2021)), mushrooms (Zulfajri et al., (2021), Pacquiao et al., (2018)), starch (Dong et al., (2020)), human nail (Chatzimitakos et al., (2018)), human hair (Singh et al., (2020)), egg (Lin et al., (2014)), over cooked meat (Wang et al., (2013)) and spider web (Sahoo et al., (2019)) are known as Green C-dots (Sharma et al., (2017)). Synthetic methods of C-dots can be classified in to top-down and bottom-up approach. In this review, we discuss about few synthetic methods of carbon dots and hetero atom doped C-dots.

SYNTHETIC METHODS FOR C-DOTS AND HETERO ATOM DOPED C-DOTS

In this section, we discuss the various approaches that can be applied for the synthesis of C-dots and hetero atom doped C-dots. In the past decades, various methods have been used for the synthesis of C-dots which include the top down and bottom approaches (as shown in figure 1).Even though these approaches play a major role in the preparation of C-dots, they have their own advantages and disadvantages. The properties of C-dots and hetero atom doped C-dots can be varied by altering the reaction conditions and methods. Based on their required applications and properties one can choose the appropriate synthetic method for C-dot preparation.

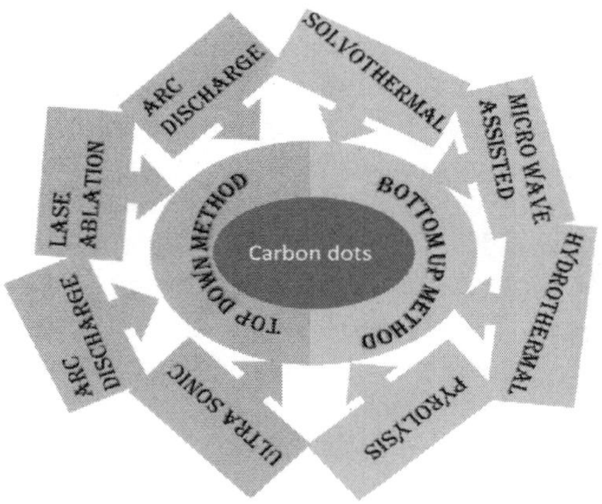

Figure1. Synthetic approaches for C-dots

TOP-DOWN APPROACH

Top-down approach denotes the formation of C-dots from the bulk material by slicing it into small particles by physical or chemical methods. The precursors used in this approach can be activated carbon, carbon nanotube, graphite, carbon black, carbon sooth, graphene oxide sheet and carbon fiber. Top-down approach includes arc discharge, laser ablation, chemical oxidation and ultrasonic method. These methods are not widely used for large scale synthesis of C-dots because of the requirement of expensive and complicated equipments, uncontrollable reaction conditions and low quantum yield.

A. Arc Discharge Method

It is the first reported synthetic method for C-dots. In this method, C-dots are prepared by passing DC current in between two graphite electrodes in the presence of argon atmosphere. Xu et al., accidently spotted C-dots for the very first time during the purification of single walled carbon nanotubes (Xu et al., (2004)). Later, Bottini et. al., synthesized fluorescent nano particles from nitric acid-oxidized carbon nano tube and pristine by this method (Bottini et al., (2006)). Even though this method produces high purity C-dots, it has disadvantages such as high cost, generation of non-uniform sized particles and requirement of high energy usage, this method is not widely used for the synthesis of C-dots(Wang et al., (2017), Ge et al., (2021)).

B. Laser Ablation Method

In laser ablation method, C-dots are prepared using a focused laser beam, which is allowed to fall on the carbon precursor in vacuum orgaseous atmosphere. Poor fluorescent C-dots of high purity along with controlled size and morphology can obtained by this method (Bottini et al., (2006), Algahtani et al., (2021), Li et al., (2021)). Sun et. al., developed C-dots for the first time by this method in the presence of water vapor and with argon as a carrier gas (Sun et al., (2006)). For sensing Hg^{2+} ions, Gonclaves et. al., synthesized C-dots by direct laser ablation (Gonçalves et al., (2010)). Subsequently, C-dots were synthesizedusing carbon nano material precursor (Li et al., (2011)). In 2011, Hu et. al., reported excitation wave length dependent fluorescent C-dots by this method (Hu et al., (2011)). Fluorescent C-dots from graphite and reduced graphene oxide were also reported in the literature (Błonski et al., (2018)).

C. Chemical Oxidation Method

Low-cost carbon precursors are treated with oxidizing agents such as nitric acid or sulfuric acid or phosphoric acidto obtain the C-dots. Even though this method do not need any complex equipment, it is not much applied for the synthesis of C-dots because of the difficulty in obtaining uniform sized particle due to harsh reaction conditions (Wang et al., (2017)). S. D Hettiarachchi et al., developed C-dots from carbon nano powder using HNO_3 and H_2SO_4 in the ratio 1:3 (Hettiarachchi et al., (2019)). C-dots with diameter 6.8 nm were prepared from coal liquefaction residue using this method (Qin et al., (2022)). Brightly fluorescent C-dots were obtained from over cooked BBQ using HNO_3(Wang et al., (2013)) Yellow fluorescence C-dots were produced from sucrose (Chang etal., (2016)). Similarly, C-dots were synthesized by oxidation of carbon nanotubes and graphite (Tao et al., 2012). S-doped C-dots and N,S-C-dots were developed from waste frying oil and human hair respectively (Hu et al., (2014), Sun et al., (2013)).

D. Ultrasonic Method

In this method C-dots are obtained by passing ultrasonic waves through the carbon precursors.Eco friendliness and short reaction time are the major attraction for this way of synthesis. A literature reported the synthesis of white fluorescent C-dots using poly amide resin precursor (Dang et al., (2016)). One pot N-doped C-dots were obtained from glucose and ammonium hydroxide in the presence of ultrasonic waves (Xu et al.,(2016)). N-doped upconversion fluorescence particles were reported by Wang et. al. (Wang et al., (2015)).

BOTTOM-UP APPROACH

Bottom-up approach can be defined as the self-assembly of smaller basic carbon units to attain a specific structure and it includes hydrothermal, solvothermal, pyrolysis and microwave assisted synthesis. This approach is widely used for the synthesis of C-dots and hetero atom doped C-dots because of its multiple advantages such as less expensiveness, usage of non-toxic precursors, less time consuming and yields particles with higher quantum yield.

A. Solvothermal Method

This method involves mixing of carbon precursor with suitable solvent in a hydrothermal vessel and carbonizing the precursor at elevated temperature to obtain the C-dots. This synthetic route is referred as a non-toxic, ecofriendly way because of the usage of biomass originated precursors. Several researchers have prepared different colored fluorescent C-dots from lemon juice (Ding et al., (2017)) mixture of tartaric acid and bran (Xu et al., (2020)) and citric acid (Javed and O'Carroll, (2021)) through this method. Similarly hetero atom doped C-dots such as Si doped C-dots and N, P co doped C-dots were also reported via this method (Dong et al., (2020), Qian et al., (2014)).

B. Hydrothermal Method

C-dots are obtained by Carbonization of the carbon precursor in a hydrothermal vessel using water as a solvent in this method. This is the one of the most effective, simplest, inexpensive, low temperature and one pot synthetic method among the top-down approaches for C-dots and hetero atom doped C-dots preparation. Similar to solvothermal, biomass assisted precursors have been mostly used for the preparation of C-dots by several researchers. C-dots were developed from apple juice (Mehta et al., (2015)), banana juice (De and Karak, (2013)), grape juice (Huang et al., (2014)) and papaya juice (Kasibabu et al., (2015)). Similar to fruit juices, fruit peels such as lemon peel (Tyagi et al., (2016)), orange peel (Prasannan and Imae, (2013)), pine apple peel (Surendran et al., (2021)), vegetables such as tomato (Miao et al., (2016)) and leaves such as Annona Reticulata leaves(John et al., (2022)), tobacco leaves(Liang et al.,(2022)) were also reported for the C-dots development. Similarly, C-dots were obtained by carbonizing the carbon precursors such as spider silk (Sahoo et al., (2019)), mushroom (Zulfajri et al.,(2021), the simultaneous production of C-dots and hydrochar (HCPacquiao et al., (2018)), human hair (Singh et al., (2020)) and shrimp egg (Lin et al., (2014)) in an autoclave. In addition to these, hetero atom doped C-dots such as N doped C-dots and N, S co-doped C-dots were formed from strawberry juice (Huang et al., (2013)) Betel leaf (Kalanidhi and Nagaraaj, (2021)) and cray fish shell (Chen et al.,(2022)) respectively.

C. Pyrolytic Method

In this method, C-dots are synthesized by carbonizing the carbon precursor at elevated temperature in the furnace. It is the one of the most simple and practicable methods for the synthesis of C-dots. However, separation of raw materials from the product formed is difficult in this method. Different fluorescent C-dots have been prepared by several researchers fromthis method. Pyrolysis was used to prepare white fluorescent C-dots from poly acrylic acid (Mao et al., (2014)) and blue emitting C- dots from sodium alginate (Fong et al., (2016)). Few reports in the literature for the synthesis of hetero atom doped C-dots by this method includes preparation of N, C-dots, N, P co-doped C-dots and green fluorescent N, S co-doped C-dots from glutamic acid, citric acid and garlic peel respectively(Das et al., (2019), Niu and Gao, (2014), Chandra et al., (2018)).

D. Microwave Assisted Method

C-dots are synthesized from carbon precursors via microwave heating within a short time by this method. Although the obtained C-dots exhibit poor reproducibility and low quality this method is a green approach for the synthesis of C-dots as it does not require any harsh chemicals. The purification of synthesized C-dots is also difficult. Both biomass originated and organic originated precursors such as quince fruit (Ramezani et al., (2018)), curry leaves (Maruthupandi et al., (2022)), and sorbitol (Mewada et al., (2014)) were used for the preparation of C-dots by several researchers. In literature, this method was also used for Boron doped C-dots and N, S co-doped C-dots from citric acid and human nail respectively (Chatzimitakos et al., (2018), Bourlinos et al., (2015)). Verities of C-dots and hetero atom doped C-dots prepared by several researchers are listed in table 1.

Table:1 Methods used to develop the C-dots and hetero atom doped C-dots

TYPE	PRECURSOR	METHOD	REFFERENCE
C-dots	Graphitic powder + Cement	Laser ablation	Sun et al., 2006
C- dots	Carbon nano powder	Chemical Oxidation	Hettiarachchi et al., 2019
C-dots	Sorbitol	Microwave assisted	Mewada et al., 2014
C- dots	Apple Juice	Hydrothermal	Mehta et al., 2015
C-dots	Lemon juice	Solvothermal	Ding et al., 2017
C- dots	Banana juice	Hydrothermal	De and Karak, 2013

TYPE	PRECURSOR	METHOD	REFFERENCE
C- dots	Grape Juice	Hydrothermal	Huang et al., 2014
N-C-dots	Strawberry Juice	Hydrothermal	Huang et al., 2013
C- dots	Lemon peel	Hydrothermal	Tyagi et al., 2016
C- dots	Orange peel	Hydrothermal	Prasannan and Imae, 2013
C- dots	Pineapple peel	Hydrothermal	Surendran et al., 2021
N, S-C-dots	Garlic peels	Pyrolysis	Das et al., 2019
C- dots	Tomato juice	Hydrothermal	Miao et al., 2016
C-dots	Curry leaves	Microwave assisted	Maruthupandi et al., 2022
C- dots	Annona Reticulata leaves	Hydrothermal	John et al., 2022
C- dots	Tobacco leaves	Hydrothermal	Liang et al., 2022
N-C-dots	Betel leaves	Hydrothermal	Kalanidhi and Nagaraaj, 2021
C- dots	Mushroom	Hydrothermal	Zulfajri et al., 2021 Pacquiao et al., 2018
N, P-C-dots	Starch	Solvothermal	Dong et al., 2020
N, S-C-dots	Human nail	Microwave assisted	Chatzimitakos et al., 2018
C- dots	Human hair	Hydrothermal	Singh et al., 2020
C- dots	Shrimp egg	Hydrothermal	Lin et al., 2014
C- dots	Over cooked BBQ	Chemical Oxidation	Wang et al., 2013
C-dots	Spider web	Hydrothermal	Sahoo et al., 2019
C- dots	Nitric-oxidized CNT & Pristine	Arc discharge	Bottini et al., 2006
C- dots	Nano-carbon-material	Laser ablation	Li et al., 2011
C- dots	Reduced Graphene oxide	Laser ablation	Błonski et al., 2018
C- dots	Coal liquefaction residue	Chemical Oxidation	Qin et al., 2022
C- dots	Sucrose+ Phosphoric acid	Chemical Oxidation	Chang et al., 2016
C- dots	Carbon nano tube + Graphite	Chemical Oxidation	Tao et al., 2012
S-C-dots	Waste frying Oil	Chemical Oxidation	Hu et al., 2014
N, S-C-dots	Human hair	Chemical Oxidation	Sun et al., 2013
C- dots	Poly amide resin	Ultrasonic treatment	Dang et al., 2016
N-C-dots	Glucose+ammonium hydroxide	Ultrasonic treatment	Xu et al., 2016
N- C-dots	Ascorbic acid	Ultrasonic treatment	Wang et al., 2015
C- dots	Tartaric acid+ Bran	Solvothermal	Xu et al., 2020
C- dots	Citric acid	Solvothermal	Javed and O'Carroll, 2021

TYPE	PRECURSOR	METHOD	REFFERENCE
Si- C-dots	Hydroquinone	Solvothermal	Qian et al., 2014
C-dots	Papaya juice	Hydrothermal	Kasibabu et al., 2015
N, S-C-dots	Cray fish shell	Hydrothermal	Chen et al., 2022
C- dots	Poly acrylic acid	Pyrolysis	Mao et al., 2014
C- dots	Sodium alginate	Pyrolysis	Fong et al., 2016
N-C-dots	Glutamic acid	Pyrolysis	Niu and Gao, 2014)
N, P-C-dots	Citric acid	Pyrolysis	Chandra et al., 2018
B-C-dots	Citric acid +Boric acid	Microwave assisted	Bourlinos et al., 2015

CONCLUSION

Carbon dots and doped carbon dots can be synthesized through top down and bottom-up approaches. While considering the advantages and disadvantages, bottom-up approaches are highly preferred for the synthesis of C-dots. Among the bottom-up methods, hydrothermal is the most extensively used approach because of its easiness and inexpensive nature. All most 70% of the reported hetero atom doped papers are about N-doped C dots. The optical property of C-dots can be varied by choosing the precursor and method of synthesis.

REFERENCES:

[1] Algahtani, A., Khan, S.H., Islam, S., Yadav, K.K., Jeon, B., 2021. Advances in the Methods for the Synthesis of Carbon Dots and their Emerging Applications., 18: 1–31

[2] Błonski, A.M.S., Gi y ska, J.C., Płocinski, M.W.T., Szymanski, L.S.Z., 2018. Fluorescent carbon and graphene oxide nanoparticles synthesized by the laser ablation in liquid. Appl. Phys. A., 124: 1–7.

[3] Bottini, M., Balasubramanian, C., Dawson, M.I., Bergamaschi, A., Bellucci, S., Mustelin, T., 2006. Isolation and characterization of fluorescent nanoparticles from pristine and oxidized electric arc-produced single-walled carbon nanotubes. J. Phys. Chem. B., 110: 831–836.

[4] Bourlinos, A.B., Trivizas, G., Karakassides, M.A., Baikousi, M., Kouloumpis, A., Gournis, D., Bakandritsos, A., Hola, K., Kozak, O., Zboril, R., Papagiannouli, I., Aloukos, P., Couris, S., 2015. Green and simple route toward boron doped carbon dots with significantly enhanced non-linear optical properties. Carbon N. Y., 83: 173–179.

[5] Chandra, S., Chowdhuri, A.R., Laha, D., Sahu, S.K., 2018. Fabrication of nitrogen- and phosphorous-doped carbon dots by the pyrolysis method for iodide and iron(III) sensing. Luminescence., 33: 336-344.

[6] Chang, M.M.F., Ginjom, I.R., Ngu-Schwemlein, M., Ng, S.M., 2016. Synthesis of yellow fluorescent carbon dots and their application to the determination of chromium(III) with selectivity improved by pH tuning. Microchim. Acta.,

183: 1899–1907.

[7] Chatzimitakos, T., Kasouni, A., Sygellou, L., Leonardos, I., Troganis, A., Stalikas, C., 2018. Human fingernails as an intriguing precursor for the synthesis of nitrogen and sulfur-doped carbon dots with strong fluorescent properties: Analytical and bioimaging applications. Sensors Actuators, B Chem., 267: 494–501.

[8] Chen, S., Ouyang, W., Zhu, Y., He, L., Zou, L., Ao, X., Liu, S., Yang, Y., Li, J., 2022. Facile Synthesis of N, S-Doped Carbon Quantum Dots from Food Waste as Fluorescent Probe for Sensitive Detection of Thiamphenicol and Its Analogues in Real Food Samples along with an Application in Bioimaging. Foods., 11: 2414.

[9] Dang, H., Huang, L.K., Zhang, Y., Wang, C.F., Chen, S., 2016. Large-Scale Ultrasonic Fabrication of White Fluorescent Carbon Dots. Ind. Eng. Chem. Res., 55: 5335–5341.

[10] Das, P., Ganguly, S., Maity, P.P., Srivastava, H.K., Bose, M., Dhara, S., Bandyopadhyay, S., Das, A.K., Banerjee, S., Das, N.C., 2019. Converting waste Allium sativum peel to nitrogen and sulphur co-doped photoluminescence carbon dots for solar conversion, cell labeling, and photobleaching diligences: A path from discarded waste to value-added products. J. Photochem. Photobiol. B Biol., 197: 111545.

[11] De, B., Karak, N., 2013. A green and facile approach for the synthesis of water soluble fluorescent carbon dots from banana juice. RSC Adv., 3: 8286–8290.

[12] Ding, H., Ji, Y., Wei, J.S., Gao, Q.Y., Zhou, Z.Y., Xiong, H.M., 2017. Facile synthesis of red-emitting carbon dots from pulp-free lemon juice for bioimaging. J. Mater. Chem. B., 5: 5272–5277.

[13] Dong, G., Lang, K., Ouyang, H., Zhang, W., Bai, L., Chen, S., Zhang, Z., Gao, Y., Mu, Z., Zhao, X., 2020. Facile synthesis of N, P-doped carbon dots from maize starchviaa solvothermal approach for the highly sensitive detection of Fe3+. RSC Adv., 10: 33483–33489.

[14] Fong, J.F.Y., Chin, S.F., Ng, S.M., 2016. A unique "turn-on" fluorescence signalling strategy for highly specific detection of ascorbic acid using carbon dots as sensing probe. Biosens. Bioelectron., 85: 844–852.

[15] Ge, G., Li, L., Wang, D., Chen, M., Zeng, Z., Xiong, W., Wu, X., Guo, C., 2021. Carbon dots: Synthesis, properties and biomedical applications. J. Mater. Chem. B., 9: 6553–6575.

[16] Gonçalves, H., Jorge, P.A.S., Fernandes, J.R.A., Esteves da Silva, J.C.G., 2010. Hg(II) sensing based on functionalized carbon dots obtained by direct laser ablation. Sensors Actuators, B Chem., 145: 702–707.

[17] Hettiarachchi, S.D., Graham, R.M., Mintz, K.J., Zhou, Y., Vanni, S., Peng, Z., Leblanc, R.M., 2019. Triple conjugated carbon dots as a nano-drug delivery model for glioblastoma brain tumors. Nanoscale., 11: 6192–6205.

[18] Hu, S., Liu, J., Yang, J., Wang, Y., Cao, S., 2011. Laser synthesis and size tailor

of carbon quantum dots. J. Nanoparticle Res., 13: 7247–7252.

[19] Hu, Y., Yang, J., Tian, J., Jia, L., Yu, J.S., 2014. Waste frying oil as a precursor for one-step synthesis of sulfur-doped carbon dots with pH-sensitive photoluminescence. Carbon N. Y., 77: 775–782.

[20] Huang, H., Lv, J.J., Zhou, D.L., Bao, N., Xu, Y., Wang, A.J., Feng, J.J., 2013. One-pot green synthesis of nitrogen-doped carbon nanoparticles as fluorescent probes for mercury ions. RSC Adv., 3: 21691–21696.

[21] Huang, H., Xu, Y., Tang, C.J., Chen, J.R., Wang, A.J., Feng, J.J., 2014. Facile and green synthesis of photoluminescent carbon nanoparticles for cellular imaging. New J. Chem., 38: 784–789.

[22] Javed, N., O'Carroll, D.M., 2021. Long-term effects of impurities on the particle size and optical emission of carbon dots. Nanoscale Adv., 3: 182–189.

[23] John, B.K., John, N., Mathew, B., 2022. Green synthesis of fluorescent carbon dots from Annona Reticulata leaves as a sensor for Chromium (III) ions. Mater. Today Proc.https://doi.org/10.1016/j.matpr.2022.06.379

[24] Kalanidhi, K., Nagaraaj, P., 2021. Facile and Green synthesis of fluorescent N-doped carbon dots from betel leaves for sensitive detection of Picric acid and Iron ion. J. Photochem. Photobiol. A Chem., 418: 113369.

[25] Kasibabu, B.S.B., D'Souza, S.L., Jha, S., Kailasa, S.K., 2015. Imaging of Bacterial and Fungal Cells Using Fluorescent Carbon Dots Prepared from Carica papaya Juice. J. Fluoresc., 25: 803–810.

[26] Li, S., Li, L., Tu, H., Zhang, H., Silvester, D.S., Banks, C.E., Zou, G., Hou, H., Ji, X., 2021. The development of carbon dots: From the perspective of materials chemistry. Mater. Today., 51: 188–207.

[27] Li, X., Wang, H., Shimizu, Y., Pyatenko, A., Kawaguchi, K., Koshizaki, N., 2011. Preparation of carbon quantum dots with tunable photoluminescence by rapid laser passivation in ordinary organic solvents. Chem. Commun., 47: 932–934.

[28] Liang, Y.M., Yang, H., Zhou, B., Chen, Y., Yang, M., Wei, K.S., Yan, X.F., Kang, C., 2022. Waste tobacco leaves derived carbon dots for tetracycline detection: Improving quantitative accuracy with the aid of chemometric model. Anal. Chim. Acta., 1191: 339269.

[29] Lin, P.Y., Hsieh, C.W., Kung, M.L., Chu, L.Y., Huang, H.J., Chen, H.T., Wu, D.C., Kuo, C.H., Hsieh, S.L., Hsieh, S., 2014. Eco-friendly synthesis of shrimp egg-derived carbon dots for fluorescent bioimaging. J. Biotechnol., 189: 114–119.

[30] Mao, L.H., Tang, W.Q., Deng, Z.Y., Liu, S.S., Wang, C.F., Chen, S., 2014. Facile access to white fluorescent carbon dots toward light-emitting devices. Ind. Eng. Chem. Res., 53: 6417–6425.

[31] Maruthupandi, M., Varatharajan, P., Shameem Banu, I.B., Hafiz Mamat, M., Vasimalai, N., 2022. White light emitting diode and anti-counterfeiting applications of microwave assisted synthesized green fluorescent carbon dots

derived from waste curry leaves. Results Opt., 8: 100249.

[32] Mehta, V.N., Jha, S., Basu, H., Singhal, R.K., Kailasa, S.K., 2015. One-step hydrothermal approach to fabricate carbon dots from apple juice for imaging of mycobacterium and fungal cells. Sensors Actuators, B Chem., 213: 434–443.

[33] Mewada, A., Pandey, S., Thakur, M., Jadhav, D., Sharon, M., 2014. Swarming carbon dots for folic acid mediated delivery of doxorubicin and biological imaging. J. Mater. Chem. B., 2: 698–705.

[34] Miao, H., Wang, L., Zhuo, Y., Zhou, Z., Yang, X., 2016. Label-free fluorimetric detection of CEA using carbon dots derived from tomato juice. Biosens. Bioelectron., 86: 83–89.

[35] Niu, J., Gao, H., 2014. Synthesis and drug detection performance of nitrogen-doped carbon dots. J. Lumin., 149: 159–162.

[36] Pacquiao, M.R., de Luna, M.D.G., Thongsai, N., Kladsomboon, S., Paoprasert, P., 2018. Highly fluorescent carbon dots from enokitake mushroom as multi-faceted optical nanomaterials for Cr 6+ and VOC detection and imaging applications. Appl. Surf. Sci., 453: 192–203.

[37] Prasannan, A., Imae, T., 2013. One-pot synthesis of fluorescent carbon dots from orange waste peels. Ind. Eng. Chem. Res., 52: 15673–15678.

[38] Qian, Z., Shan, X., Chai, L., Chen, J., Feng, H., 2014. Si-Doped Carbon Quantum Dots: A Facile and General Preparation Strategy, Bioimaging Application, and Multifunctional Sensor. ACS Appl. Mater. Interfaces., 6: 6797 6805.

[39] Qin, F., Li, Q., Tang, T., Zhu, J., Gan, X., Chen, Y., Li, Y., Zhang, S., Huang, X., Jia, D., 2022. Functional carbon dots from a mild oxidation of coal liquefaction residue. Fuel., 322: 124216.

[40] Radnia, F., Mohajeri, N., Zarghami, N., 2020. New insight into the engineering of green carbon dots: Possible applications in emerging cancer theranostics. Talanta., 209: 120547.

[41] Ramezani, Z., Qorbanpour, M., Rahbar, N., 2018. Green synthesis of carbon quantum dots using quince fruit (Cydonia oblonga) powder as carbon precursor: Application in cell imaging and As3+ determination. Colloids Surfaces A Physicochem. Eng. Asp., 549: 58–66.

[42] Sahoo, N.K., Das, S., Jana, G.C., Aktara, M.N., Patra, A., Maji, A., Beg, M., Jha, P.K., Hossain, M., 2019. Eco-friendly synthesis of a highly fluorescent carbon dots from spider silk and its application towards Hg (II) ions detection in real sample and living cells. Microchem. J., 144: 479–488.

[43] Sharma, V., Tiwari, P., Mobin, S.M., 2017. Sustainable carbon-dots: Recent advances in green carbon dots for sensing and bioimaging. J. Mater. Chem. B., 5: 8904–8924.

[44] Singh, A., Eftekhari, E., Scott, J., Kaur, J., Yambem, S., Leusch, F., Wellings, R., Gould, T., Ostrikov, K., Sonar, P., Li, Q., 2020. Carbon dots derived from human hair for ppb level chloroform sensing in water. Sustain. Mater. Technol., 25: e00159.

[45] Sun, D., Ban, R., Zhang, P.H., Wu, G.H., Zhang, J.R., Zhu, J.J., 2013. Hair fiber as a precursor for synthesizing of sulfur- and nitrogen-co-doped carbon dots with tunable luminescence properties. Carbon., 64: 424–434.

[46] Sun, Y.P., Zhou, B., Lin, Y., Wang, W., Fernando, K.A.S., Pathak, P., Meziani, M.J., Harruff, B.A., Wang, X., Wang, H., Luo, P.G., Yang, H., Kose, M.E., Chen, B., Veca, L.M., Xie, S.Y., 2006. Quantum-sized carbon dots for bright and colorful photoluminescence. J. Am. Chem. Soc., 128: 7756–7757.

[47] Surendran, P., Lakshmanan, A., Priya, S.S., Geetha, P., Rameshkumar, P., Kannan, K., Hegde, T.A., Vinitha, G., 2021. Fluorescent carbon quantum dots from Ananas comosus waste peels: A promising material for NLO behaviour, antibacterial, and antioxidant activities. Inorg. Chem. Commun., 124: 108397.

[48] Tao, H., Yang, K., Ma, Z., Wan, J., Zhang, Y., Kang, Z., Liu, Z., 2012. In vivo NIR fluorescence imaging, biodistribution, and toxicology of photoluminescent carbon dots produced from carbon nanotubes and graphite. Small., 8: 281–290.

[49] Tyagi, A., Tripathi, K.M., Singh, N., Choudhary, S., Gupta, R.K., 2016. Green synthesis of carbon quantum dots from lemon peel waste: Applications in sensing and photocatalysis. RSC Adv., 6: 72423–72432.

[50] Wang, F., Wang, S., Sun, Z., Zhu, H., 2015. Study on ultrasonic single-step synthesis and optical properties of nitrogen-doped carbon fluorescent quantum dots. Fullerenes Nanotub. Carbon Nanostructures., 23: 769–776.

[51] Wang, J., Sahu, S., Sonkar, S.K., Tackett, K.N., Sun, K.W., Liu, Y., Maimaiti, H., Anilkumar, P., Sun, Y.P., 2013. Versatility with carbon dots-from overcooked BBQ to brightly fluorescent agents and photocatalysts. RSC Adv., 3: 15604–15607.

[52] Wang, R., Lu, K.Q., Tang, Z.R., Xu, Y.J., 2017. Recent progress in carbon quantum dots: synthesis, properties and applications in photocatalysis. J. Mater. Chem. A., 5: 3717–3734.

[53] Wang, Y., Li, X., Zhao, S., Wang, B., Song, X., Xiao, J., Lan, M., 2022. Synthesis strategies, luminescence mechanisms, and biomedical applications of near-infrared fluorescent carbon dots. Coord. Chem. Rev., 470: 214703.

[54] Xu, J., Wang, C., Li, H., Zhao, W., 2020. Synthesis of green-emitting carbon quantum dots with double carbon sources and their application as a fluorescent probe for selective detection of Cu2+ ions. RSC Adv., 10: 2536–2544.

[55] Xu, Q., Kuang, T., Liu, Y., Cai, L., Peng, X., Sreenivasan Sreeprasad, T., Zhao, P., Yu, Z., Li, N., 2016. Heteroatom-doped carbon dots: synthesis, characterization, properties, photoluminescence mechanism and biological applications. J. Mater. Chem. B., 4: 7204–7219.

[56] Xu, X., Ray, R., Gu, Y., Ploehn, H.J., Gearheart, L., Raker, K., Scrivens, W.A., 2004. Electrophoretic analysis and purification of fluorescent single-walled carbon nanotube fragments. J. Am. Chem. Soc., 126: 12736–12737.

[57] Zhang, J., Yu, S., 2016. Carbon dots : large-scale synthesis , sensing and bioimaging. Biochem. Pharmacol., 19: 382–393.

[58] Zulfajri, M., Kao, Y.T., Huang, G.G., 2021. Retrieve of residual waste of carbon dots derived from straw mushroom as a hydrochar for the removal of organic dyes from aqueous solutions. Sustain. Chem. Pharm., 22: 100469.

Chapter 10

Formic Acid, A Convenient Hydrogencarrier, and its Dehydrogenation by Palladium

N Sai Kumar[1], Adhikary, A[1*]
[1]*VIT-AP University, Amaravati, Andhra Pradesh, India*
E-mail: anubendu.a@vitap.ac.in

INTRODUCTION

Hydrogen economy is in high demand due to the limited reserve of fossil fuels and its negative impact on the environment. Dihydrogen (H_2), a non-polluting propellant, can directly be combusted to produce heat ($\Delta H_{combustion}$ = 286 kJ/mol), and, more importantly, it can be fed into fuel cells to produce clean electricity. In addition, dihydrogen can be produced in large scales from renewable sources through the steam reforming process of natural gas, and from solar powered water splitting, which makes it a promising future fuel.[1]However, because of its gaseous state (at temperature >-253 °C, under atmospheric pressure) and extreme flammability, H_2-storage and transportation is always a challenge. In industry, present techniques for H_2 storage include liquefaction and compression, which are energy intensive, costly and, not to mention, risky. A convenient substitute will be appropriate liquid organic hydrogen carriers (LOHC), which can be used in existing oil transportation infrastructure.[2]

FORMIC ACID FOR HYDROGEN STORAGE AND TRANSPORTATION

In the past two decades, several LOHC materials have been investigated for effectively and safely storing-carrying H_2, and undoubtedly, formic acid is one of the most convenient and promising materials among them. Formic acid has a high volumetric (53.4 g/L) and weight (4.4 wt%) H_2 capacity. It is highly soluble in water, and its aqueous solution is non-flammable and non-toxic.[3]Most importantly, chemically converting H_2 into FA on a large scale is a well-developed economically viable process such as CO_2 hydrogenation. There are several efficient catalytic systems for producing formic acid by hydrogenating CO_2, an inexpensive, nontoxic feedstock (figure 1).[4]

$$\text{Dehydrogenation: } HCOOH \rightarrow H_2 + CO_2 \qquad \Delta G = -48.4 \text{ kJ/mol} \quad (1)$$

$$\text{Dehydration: } HCOOH \rightarrow H_2O + CO \qquad \Delta G = -28.5 \text{ kJ/mol} \quad (2)$$

Figure 1: Two pathways of formic acid decomposition

Hydrogen, stored in formic acid, can be released through chemical decomposition. There are two possible decomposition pathways as shown in figure 1: dehydrogenation (1) and dehydration (2). Both of these transformations are associated with high activation energy, and therefore, need catalysts. For the purpose of on-demand release of hydrogen gas, the formic acid dehydrogenation (FAD) process has to be selective towards dehydrogenation over dehydration. On the other hand, the later path results in the formation of carbon monoxide (CO), which is extremely harmful (<10 ppm) to fuel cell catalysts.[5] Therefore, it is highly desirable to develop methods for decomposition of FA selectively to H_2 and CO_2. For the last 10-15 years extensive research has been done on various catalytic systems under homogeneous and heterogeneous conditions, and significant advancements have been made to generate CO-free H_2.[6]

SOLID CATALYSTS FOR HYDROGEN PRODUCTION

Over the past decades, applicability of transition metal-based catalysts under heterogeneous conditions have been widely studied. Interestingly, it is observed that there are several components of these nanomaterial catalysts, which are critical towards the reactivity as well as in determining selectivity. The electronic nature of the metal surface as well as the support and their interaction type are crucial. In recent years, immense progress towards efficient FA-dehydrogenation catalysis has been made through the variation of metal nanoparticles' (NP) size, use of numerous supporting materials, following different synthetic strategies, altering compositions of catalysts etc.[7]Successfully, a number of solid catalysts have been developed which can selectively produce CO-free H_2 from FA. However, there are rooms for improvement especially in terms of efficiency, avoiding additives and promoters, and in lowering the overall cost.

UNIQUE PROPERTY OF PALLADIUM

In earlier periods, activity of transition metals and their oxides were studied in vapor phase formic acid decomposition (FAD), which involves a high temperature. Achieving a robust reaction condition and presence of formic acid vapor in produced gasses will be challenging for fuel cell use. To develop low temperature procedures in aqueous media, theoretical and thorough experimental studies have been done by several research groups focusing on the crucial role of various catalytic components especially transition metals and support materials. To find the suitable metal for HCO_2H activation, though a number of theoretical calculations are performed concentrating on the reactivity of transition metals towards similar small molecules such as CO, CO_2, H_2 etc., unfortunately, only a few studies have been reported on metal-formic acid interaction. DFT study by Mavrikakis group found that activity of metal surfaces with FA follows two opposite trends for the two individual steps of H-atoms dehydrogenation (H_1 and H_2), from FA ($H_{1/2}COOH_{2/1}$).[8]This study, based on binding energy (of reactants and intermediates) calculation of transition metal-surfaces, predicts that in 1st H-atom dehydrogenation step there is an increase in activity from right to left in the periodic table i.e. more facile for Re, Ru, Os. Conversely, the 2nd dehydrogenation step is more favorable for transition metals on the right side of the periodic table (e.g. Cu, Ag, Au). Hence, the optimal reactivity will be expectedfrom group 10 transition metals, i.e. Pt and Pd. An experimental study by Solymosi group on vapor phase FA decomposition with metal catalysts, supported on inactive carbon (Norrit), produced the following reactivity order: Ir > Pt > Rh > Pd > Ru.[9]Tsang and co-workers compared catalytic activity of several metal colloids through investigating ^{13}C NMR spectroscopy.[10]The observed order of initial FA-decomposition rate is as follows: Pd > Rh > Pt Ru > Au > Ag, showing the preferability for palladium in FAD.

CATALYSTS FOR FORMIC ACID DEHYDROGENATION

These comparative analyses are not fully consistent with each other, nevertheless, palladium is mostly studied and found to be the most effective metal in producing H_2 from FA. Diverse techniques have been implied to synthesize palladium nanoparticles (Pd-NP) on various supports and their catalytic reactivity in FAD have been explored. Some of the selected catalysts are shown intable 1.

Table 1. Heterogeneous palladium catalysts in formic acid dehydrogenation

S.No.	Metal	Support	Reference
1.	Pd/C (10%)	Heraeus.co	11
2.	Pd/$CN_{0.25}$	Chitosan	12
3.	Pd/C-m (6%)	Vulcan XC-72	13
4.	Pd-B/C	Vulcan XC-72	14
5.	Pd/C-ca	Vulcan XC-72	15
6.	Pd/C (20%)	Vulcan XC-72	16
7.	Pd/C (5%)	Sigma-Aldrich	17
8.	Pd/C (5%)	Charcoal	18
9.	Pd/C (5%)	Vulcan XC-72	19
10.	Pd/C	Vulcan XC-72	20
11.	Pd/C (0.5%)	ShirasagiM	21
12.	Pd/C (10%)	Heraeus.co	22
13.	Pd-MIL-125	MOF: Ti and 1,4-benzene dicarboxylic acid	23
14.	Pd-NH_2-MIL-125	MOF: Ti and 2-amino 1,4-benzene dicarboxylic acid	23
15.	Pd-NH_2-MIL-125i	MOF: Ti and 2-amino 1,4-benzene dicarboxylic acid	23
16.	Pd-B/C	Supported on Vulcan XC-72	24

17.	Pd-C-$NaBH_4$	Supported on Vulcan XC-72	25
18.	Pd-C-HCOONa	Supported on Vulcan XC-72	25
19.	Pd/MSC-30	Supported on MSC-30 (prepared with NaOH)	26
20.	Pd/MSC-30	Supported on MSC-30 (prepared w/o NaOH)	26

LIMITATION AND SCOPE

Although reactivity wise the role of palladium is valuable, Pd (0) nanoparticles have a very high affinity towards carbon monoxide, the unwanted dehydration product (eqn. 2). Therefore, low selectivity, hence, CO-poisoning could be disadvantageous for palladium in FAD. In fact, CO-adsorption along with catalyst poisoning were observed for many monometallic Pd NP catalysts, and undoubtedly a serious concern for further catalyst modification. Following this direction, use of transition metals having low CO-affinity were attempted. After palladium, gold is found to be the second most effective transition metal in this reaction, which has a very low binding energy with CO. Catalytic reactivity of other monometallic NPs such as Au, has also been investigated thoroughly.[27-33]Furthermore, bimetallic alloys, core-shell structure, and a few trimetallic systems have also been explored in search of higher reactivity and improved selectivity towards formic acid dehydrogenation.

PALLADIUM IN FAD UNDER HOMOGENEOUS CONDITION

Although Pd based catalysts are found to be most active and applied under heterogeneous condition, by our knowledge there is no well-defined palladium complex applied in this reaction. Therefore, it is highly desirable to study various palladium complexes in homogeneous formic acid dehydrogenation.

CONCLUSION

Undoubtedly, till now palladium is mostly studied as an active metal in producing hydrogen from formic acid. However, further improvement of the catalysts is required with respect to efficiency, selectivity, and economy. Hence, detailed and intense research is required in this field. Currently, the directions of progress are towards a) preparing and applying bimetallic or trimetallic Pd catalysts, and b) use of monometallic but less-costly metals other than Pd. Additionally, palladium complexes with a variety of ligands need to be studied to explore their reactivity under the homogeneous formic acid decomposition process.

REFERENCES

[1] 2016 Annual Progress Report, 2017, Hydrogen and Fuel Cells Program, Department of Energy, USA

[2] Phillimon.M.Modisha,Rudaviro garidzirai,Peter wasserschied, Dmitri Bessarabov, 2019. The prospect of hydrogen storage using liquid organic hydrogen carrier, Energy Fuels, 33, 4, 2778–2796

[3] Jorg Eppinger, Kuo-Weihuang, 2017. Formic acid as a hydrogen energy carrier. ACS Energy Lett., 2, 1, 188–195

[4] Ruiyan Sun, Yuhe Liao, Shao-tao Bai, Mingyuan zheng, Cheng Zhou, Tao Zhang and Bert F.sels,2021. Heterogeneous catalyst for CO2 hydrogenation to formic acid/Formate, From nanoscale to single atom Energy Environ. Sci. 14, 1247-1285

[5] Zhingang Qi, Chunzhi He, Arthur Kaufman, 2002. Effect of CO in the anode fuel on the performance of PEM fuel cell Cathode. Journal of Power Sources 111 2 : 239-247.

[6] Naoya Onishi, Ryoichi Kanega, Hajime Kawanami, Yuichiro Himeda. 2022. Recent progress in homogeneous catalytic dehydrogenation of formic acid. Molecules. Jan 11;27 2 : 455.

[7] Gregory J Kubas. 2007, Fundamentals of H_2 binding and reactivity on transition metals undergoing hydrogenase function and H_2 production and storage. Chem. Rev.107, 4152 4205.

[8] Jeffrey A Herron, Jessica Scaranto, Peter Ferrin, Shali, Manos Mavrikakis. 2014. Trends in fromic acid decomposition on model transition metals surface: A density functional theory study. ACS Catal., 4, 4434 4445.

[9] F. Solymosi. A.Koose, N.Lilliom, and I.Ugrai. 2011. Production of CO-free H2 from formic acid, A comparative study of the catalytic behavior of Pt metals on a carbon support. Journal of Catalysis., 279: 213–219.

[10] Simon Jones, Jin Qu,Dr. Karaked Tedsree, Prof. Xue-Qing Gong, Prof. Shik Chi Edman TsangAngew. 2012. Prominent electronic and geometric modifications of palladium nanoparticles by polymer stabilizers for hydrogen production under ambient conditions. Chem. Int. Ed., 51, 11275-11278.

[11] Hu, C. Pulleri, J. K. Ting, S.-W and Chan, K.Y. 2014. Activity of Pd/C for hydrogen generation in formic acid solution. Int. J. Hydrogen Energ., 39: 381–390.

[12] Bi, Q.Y. Lin, J.D. Liu, Y.M. He, H.Y. Huang, F.Q. and Cao, Y. 2016. Gold supported on zirconia polymorphs for hydrogen generation from formic acid in base-free aqueous medium Angew. Chem. Int. Ed. 55: 1–6.

[13] Zhu, Q.L. Tsumori, and N. Xu, Q. 2015. Immobilizing Extremely Catalytically Active Palladium Nanoparticles to Carbon Nanospheres: A Weakly-Capping Growth Approach J. Am. Chem. Soc., 137: 11743–11748.

[14] Yang, Z.L. Yan, J.M.; Wang, H.-L.; Ping, Y. and Jiang, Q. 2012. Pd/C synthesized with citric acid: An efficient catalyst for hydrogen generation from acid/sodium formate. Sci. Rep., 2: 598.

[15] Lv, Q. Feng, L. Hu, C. Liu, C. and Xing, W. 2015. High-quality hydrogen generated from formic acid triggered by in situ prepared Pd/C catalyst for fuel cells. Catal. Sci. Technol., 5: 2581–2584.

[16] Sanchez, F. Motta, D. Roldan, A. Hammond, C. Villa, and A. Dimitratos, N. 2018. Hydrogen generation from additive-free formic acid decomposition under mild conditions by Pd/C: Experimental and DFT studies Top Catal., 61: 254–266.

[17] Lv, Q. Meng, Q. Liu, W. Sun, N. Jiang, K. Ma, L. Peng, Z. Cai, W.-B. Liu, C. Ge, J. Liu, L.-M. and Xing, W.2018. Pd-PdO Interface as active site for HCOOH selective dehydrogenation at ambient condition. J. Phys. Chem. C., 122: 2081–2088.

[18] Zhang, S. Jiang, B. Jiang, K. Cai, and W.-B. 2017. Surfactant-free synthesis of carbon-supported palladium nanoparticles and size-dependent hydrogen production from formic acid-formate solution. ACS Appl. Mater. Interfaces, 9: 24678–24687.

[19] Wang, X. Q, G.-W. Tan, C.-H. Li, Y.-P. Guo, J. Pang, X.-J. Zhang, and S.Y. 2014. Pd/C nanocatalyst with high turnover frequency for hydrogen generation from the formic acid-formate mixtures. Int. J. Hydrogen Energ. 39: 837–843.

[20] Navlani-Garcia, M. Mori, K. Wen, M. Kuwahara, Y. and Yamashita, H. 2015. Size effect of carbon-supported Pd nanoparticles in the hydrogen production from formic acid. Bull. Chem. Soc. Jpn., 88: 1500–1502.

[21] Navlani-Garcia, M.; Mori, K.; Wen, M.; Kuwahara, Y.; Yamashita, H. *Bull. Chem. Soc. Jpn.*__2015__,*88*, 1500–1502.

[22] Mori, K.; Dojo, M.; Yamashita, H. *ACS Catalysis*__2013__, *3*, 1114–1119.

[23] Martis, M. Mori, K. Fujiwara, K. Ahn, W.-S. And Yamashita, H. 2013. Amine functionalized MIL-125 with imbedded palladium Nanoparticles as an efficient catalyst for dehydrogenation of formic acid at ambient temperature. J. Phys. Chem. C., 117: 22805–22810.

[24] Jiang, K. Xu, K. Zou, S. Cai, and W.B. 2014. B-Doped Pd catalyst: boosting room-temperature hydrogen production from formic acid-formate solutions J. Am. Chem. Soc., 136: 4861–4864.

[25] Zhu, Q.-L.; Tsumori, N.; Xu, Q. 2015. Immobilizing extremely catalytically active palladium nanoparticles to carbon nanospheres. A weakly capping growth approach. J. Am. Chem. Soc., 137: 11743–11748.

[26] Zhu, Q.L. Tsumori, and N. Xu, Q. 2014. Sodium hydroxide assisted growth of uniform palladium nanoparticles on nanoporous carbon MSC-30 for efficient and completed dehydrogenation of formic acid under ambient conditions. Chem. Sci., 5: 195–199.

[27] Qing-Yuan Bi, Xian-Long Du, Yong-Mei Liu, Yong Cao, He-Yong He, and Kang-Nian Fan. 2012. Efficient subnanometric gold-catalyzed hydrogen generation via formic acid decomposition under ambient conditions., J. Am. Chem. Soc., 134, 8926–8933.

[28] Qinggang Liu, Xiaofeng Yang, Yanqiang Huang, Shutao Xu, Xiong Su, Xiaoli Pan, Jinming Xu, Aiqin Wang, Changhai Liang, Xinkui Wang b and Tao Zhang. 2015. A Schiff Base Modified Gold Catalyst for Green and Efficient H2 Production from Formic Acid. Energy Environ. Sci., 8, 3204–3207.

[29] Ojeda, M. Iglesia, E. 2009. Formic Acid Dehydrogenation on Au-Based Catalysts at NearAmbient Temperatures. Angew. Chem. Int. Ed., 48, 4800–4803.

[30] Yadav, M. Akita, T. Tsumori, N. Xu, Q. J. 2012. Amine-fleshed gold nanoparticles encapsulated in silica nanospheres highly active for catalytic decomposition of formic acid Mater. Chem. 22, 12582–12586.

[31] Bi, Q.-Y. Lin, J.-D. Liu, Y.-M. He, H.-Y. Huang, F.-Q. Cao, Y. 2016. Gold supported on zirconia polymorphs for hydrogen generation from formic acid in base-free aqueous medium J. Power Sources., 328, 463–471.

[32] Mielby, J.; Kunov-Krause, A. J.; Kegnaes, S. 2017. Decomposition of formic acid over silica encapsulated and amine functionalized gold nanoparticles. J. Cat., 345, 149–156.

[33] Bulushev, D. A. Beloshapkin, S. Ross, J. R. H. 2010. Production of Hydrogen: Photocatalytic Decomposition of Dimethyl Ether over Metal-Promoted TiO2 Catalysts Catal. Today., 154, 7–12.

Chapter 11

A Study of Carbon Nanotube FETs with Four Different Gate Insulators

Md. Akram Ahmad[1], Jitendra Kumar[1] and Bhubon Mech[2]
[1]*Indian Institute of Technology, Dhanbad, 826004, India*
[2]*Defence Institute of Advanced Technology, Pune, 411025, India*
E-mail: akram14407@gmail.com

INTRODUCTION

Modern technological development has forced device dimensions into the nanometer region, which is the ultimate way to improve the performance of digital circuits. However, sinking the size of CMOS devices affects challenges like drain-induced barrier lowering (DIBL), leakage current, subthreshold slope (SS), and velocity saturation; therefore, retaining Moore's Law is challenging due to several limitations Si-based CMOS devices. Therefore, researchers are looking for alternate devices to replace the existing silicon-based technology [1-3]. Carbon nanotubes (CNTs) have become a strong candidate for proficient integration through silicon-based devices because of their unique features. CNT agreements high mobility, fast response, high-drive current capability and velocity saturation, resulting in more excellent ON-current (I_{ON}) in CNT-based devices related to CMOS devices [4-6]. Moreover, the variable bandgap of CNT with a lesser diameter reveals less OFF-current (I_{OFF}) [7]. Hence, CNT offers a high current ON/OFF ratio (I_{ON}/ I_{OFF}) without suffering short-channel effects (SCEs), which are grievous in conventional metal oxide semiconductor field-effect transistors (MOSFET).

CNT can be used as a channel material in a field-effect transistor (FET) to form CNTFET, and CNT as channel material with metal source/drain contacts FETs is known as Schottky barrier CNTFETs (SBCNTFETs) [8]. The performance of SBCNTFET is decreased due to ambipolar currents; however, the device's performance can be enhanced by introducing doping in the CNT, which expands the I_{ON}/ I_{OFF} ratio and subthreshold swing (SS) [9]. In recent years, much research has been done to reduce leakage current [10-12] to recognize low-power circuit applications. In [13], a novel structure of CNTFET is offered, showing that CNTFETs are possible for low-power circuit design. The high-k structure also causes further immunity against short-channel effects, including drain-induced barrier-lowering and subthreshold swing [14]; the results also demonstrate that a very high I_{ON}/I_{OFF} ratio can be possible due to higher saturation and lower leakage currents.

The goal of this report is to examine the effects of four-different gate insulator (FDGI) in the CNTFET device structure. The digital performance parameter, I_{ON}/I_{OFF}, of CNTFET with FDGI engineering is evaluated and compared with conventional Low-k CNTFET, which has a constant dielectric material of gate insulator of the device. The device structure is simulated in open-source atomistic device simulator NanoTCAD ViDES using the non-equilibrium Green's function (NEGF) method.

DEVICE STRUCTURE AND SIMULATION METHODOLOGY

Fig. 1 shows the simulated double-gated device structure, consisting of CNT as the channel material with chirality of (13, 0), having a band gap of 0.75 eV. The parameters considered in the simulation are gate length (L_G) of 5 nm, drain and source length (L_D and L_S) of 5 nm are the extension of gate region. The doping concentration of donor-type source and drain reservoirs is 2.5 10^{13} cm^{-2}. The physical thickness of gate insulator is of 2 nm, and the drain is biased with a voltage of 0.5 V at a temperature of 300 K.

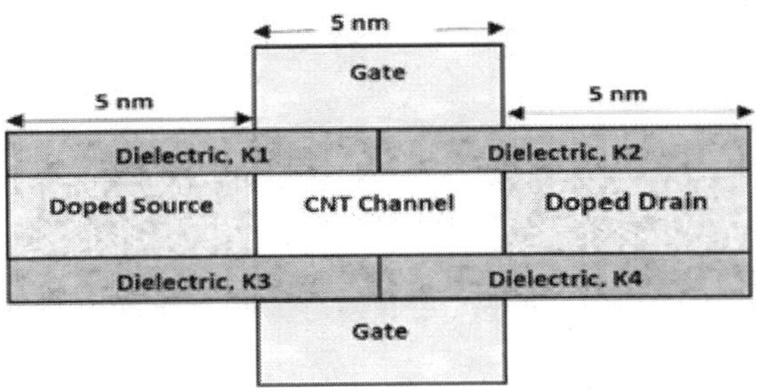

Fig. 1. The simulated FDGI- CNTFET structure with doped source and drain extension.

The simulations of CNT FETs have been performed under the non-equilibrium Green's function (NEGF) method with a tight-binding model [14], and NEGF is solved self-consistently with 3-D Poison's equation. At first, the model has to identify an appropriate Hamiltonian matrix for the channel [14] in real space, and then retarted Green's function is obtained from the equation:

$$G(E) = \left[EI - H - \sum_S - \sum_D\right]$$

where E, I, H, Σ_s and Σ_s are the energy, identity matrix, Hamiltonian of CNT, self-energy matrix of the source, and self-energy of drain, respectively. The first nearest neighbor tight-binding approximation specifies the atomic collaboration among carbon-carbon (C-C) atoms.

Once *G(E)* is calculated, then hole and electron concentrations can be calculated as [15] by solving the Schrodinger equation using the open boundary condition.

After that, the Newton-Raphson iteration method is applied to get electrostatic potential self-consistently. Finally, drain current is attained using the Landauer formula, as discussed in [15].

RESULTS AND DISCUSSION

Using the methodology described above, the performance evaluation of CNTFET with different dielectric materials of gate insulator is discussed in below section:

A. FDGI-CNTFET with K = 16 and K = 25

The first FDGI-CNTFET structure considered is the device having dielectric constants K= 16 and K= 25, for all of four sections of gate insulator which is compared with the conventional structure having K = 3.9 for all of the sections of gate insulator in both layers.

At first, the transfer characteristics (I_D-V_{GS}) of the device with K= 16 and 25 in all of four sections of gate insulator are simulated and compared with the conventional low-k CNTFET (K = 3.9) as shown in Fig. 2(a). It is worth mentioning that the minimum current is considered as the I_{OFF}; Also, the maximum current of the device is considered as I_{ON}. It is observed from Fig. 2 (a) that FDGI CNTFET with K = 16 and 25 in all of four sections of gate insulator has a lower OFF-current and higher ON-current in comparison with the conventional low-k CNTFET with K = 3.9. The increase in dielectric constant not only does decrease OFF-current but also I_{ON} increases, as depicts in Fig. 2(a). It is because of, when dielectric constant increases, the conduction band profile decreases as shown in Fig. 2(b), whereas decreasing transmission coefficient, in Fig 2(c) and Fig. 2(d), led to decrease I_{OFF}. In reality, when dielectric constant rises, the band-to-band tunnelling currents shrink and thermionic emission current, which is the thermally induced flow of the charge carriers from a potential-energy barrier growths. Instead, rising thermionic emission current leads to greater I_{ON}. In addition, reducing band-to-band tunnelling current leads to lower I_{OFF}. Thus, I_{ON}/ I_{OFF} ratio is increased [16, 17].

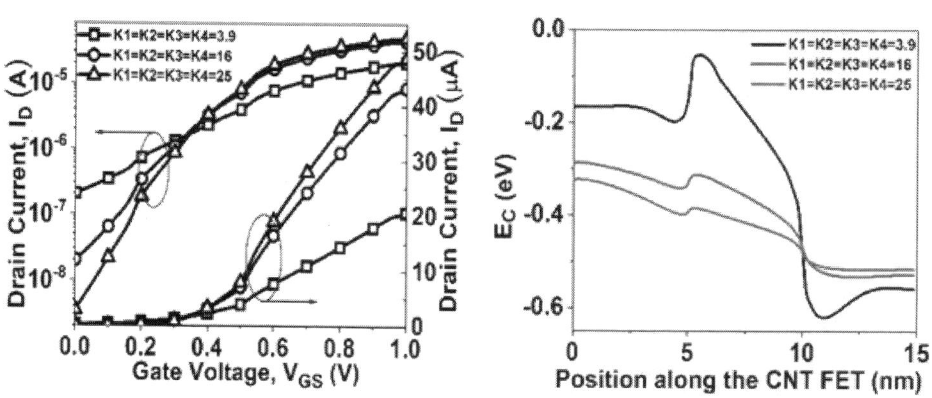

(a) (b)

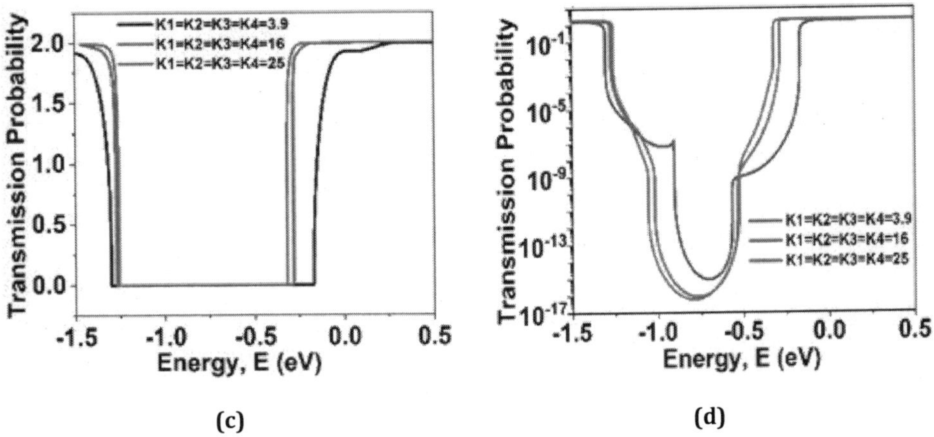

Fig. 2. (a) Transfer characteristics of CNT FET at V_{DS} =0.5 V (b) CB profile of the CNT FET at ON state (c) Variation of transmission probability with energy in linear-scale (d) Variation of transmission probability with energy in log-scale.

B. FDGI-CNTFET with K1=K2=25 and K3=K4=3.9

The second FDGI-CNTFET structure proposed is the device having dielectric constants K1= K2=25 in the right and left of top layer and also K3=K4=3.9 in the right and left of bottom layer of gate insulator which is compared with the device having K=3.9 for all of the sections of gate insulator in both layers.

Fig. 3(a) depicts the I_D-V_{GS} of the device CNTFET with K = 3.9 and FDGI-CNTFET with dielectric constant K3 = K4 = 3.9 in the right and left of the bottom layer of gate insulator and K1 = K2 = 25 in the right and left of the top layer are simulated and compared to each other. From Fig. 3 (a), it is observed that the proposed FDGI-CNTFET with K1 = K2 = 25 and K3 = K4 = 3.9 of gate insulator has a higher I_{ON} and lower I_{OFF} in comparison with the conventional CNTFET (K = 3.9). Consequently, with a high dielectric material only in the top layer is led to more reduction in the band-to-band tunnelling current and I_{OFF} decreases more than the conventional one as well as the previous structure with K = 25 for all of the sections of gate insulator. Furthermore, I_{ON} is not improved very much in contrast with the earlier structure with K = 25 for all of the sections of gate insulator. Fig. 3(b) shows conduction band profile of simulated structure, which shows larger bandgap is for the conventional low-k CNTFET compared to the proposed structure, which results decrease in OFF-current and increase in on-current of the proposed device compared to the conventional CNTFET. It is also observed from Fig. 3 (c) and Fig. 3 (d) that there is reduction in transmission probability graph for the proposed structure, which results lower OFF-current compared to the conventional CNTFET.

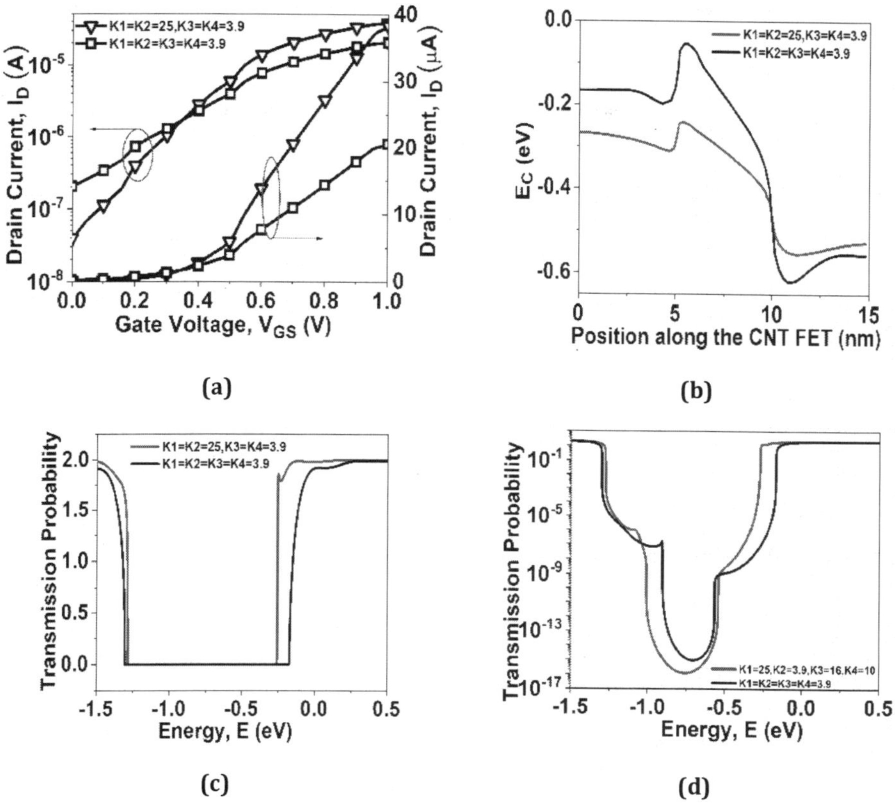

Figure 3. (a) Transfer characteristics of CNT FET at V_{DS} =0.5 V (b) CB profile of the CNT FET at ON state (c) Variation of transmission probability with energy in linear-scale (d) Variation of transmission probability with energy in log-scale

C. FDGI-CNTFET with K1=25, K2= 3.9, K3=16 and K4=10

The third FDGI-CNTFETs structure proposed is the device having dielectric constants K1 = 25, K2 = 3.9, K3 = 16 and K4 = 10 in the top and bottom layer of gate insulator which is compared to the conventional low-k CNTFET with K = 3.9 for the whole sections of gate insulator in both layers.

Fig. 4(a) depicts the I_D-V_{GS} of the device CNTFET with conventional low-k CNTFET and FDGI-CNTFET with dielectric constant K1 =25, K2 = 3.9 in the right and left of the top layer and K3 = 16, K4 = 10 in the right and left of the bottom layer of gate insulator. It is observed from the Fig. 4 (a), the proposed FDGI-CNTFET with K1=25, K2 = 3.9 in the right and left of the top layer and also K3 = 16, K4 = 10 in the right and left of the bottom layer of gate insulator has a low I_{OFF} and high I_{ON} in compare to the conventional low-k CNTFET with K = 3.9. Fig. 4(b) describes the conduction band profile of the conventional low-k CNTFET (K = 3.9) as well as FDGI-CNTFET with K1 = 25, K2 = 3.9 in the right and left of the top layer and also K3 = 16, K4 = 10 in the right and left of the bottom layer of gate insulator. It is observed from the Fig. 4(b) that lower potential barrier near the source for proposed

FDGI-CNTFET in comparison with the conventional CNTFET. Actually, in both layers of the proposed structure with high dielectric constant 25 and16 at the source side increases I_{ON} and decreases the I_{OFF}, whereas device with low dielectric constant 3.9 and 10 at the drain side forms smaller capacitances [16, 17]. Hence, in proposed structure, the transmission probability decreases which results in a low OFF-current compared to the conventional one, which is shown in Fig. 4 (c) and Fig. 4 (d).

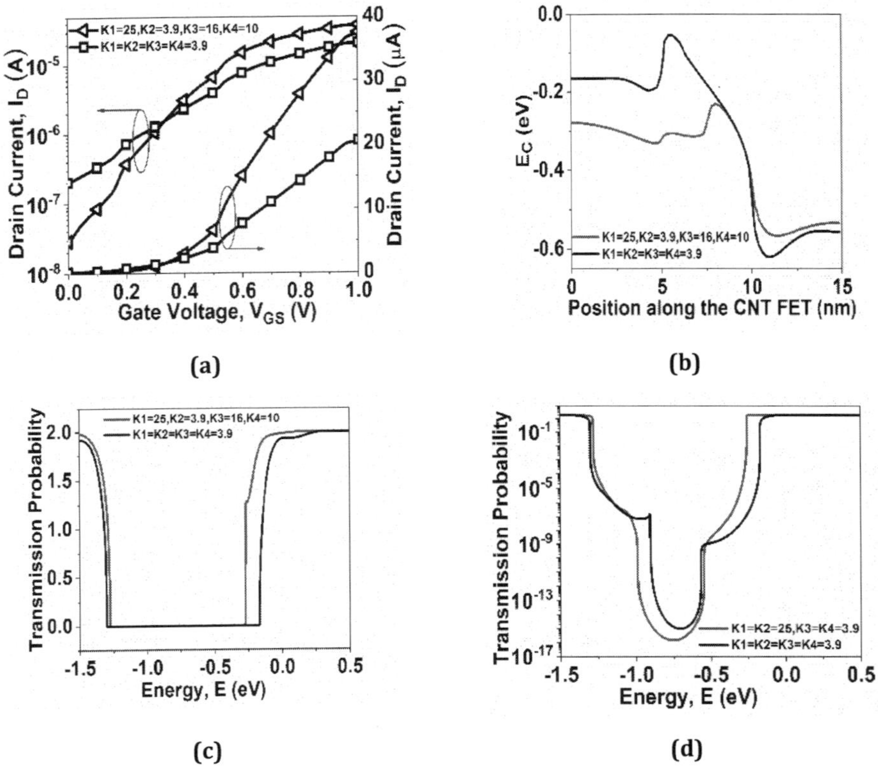

Fig. 4. (a) Transfer characteristics of CNT FET at V_{DS} =0.5 V (b) CB profile of the CNT FET at ON state (c) Variation of transmission probability with energy in linear-scale (d) Variation of transmission probability with energy in log-scale.

D. FDGI-CNTFET with K1= K3= 25, K2= K4= 3.9, and K1=K3=3.9, K2=K4=25

The forth FDGI-CNTFET proposed structure is introduced in two states, which are shown in below figure.

Out of the two proposed states, one of them is involved with dielectric constants K1=K3=25, K2=K4=3.9 in the top and bottom layer of gate insulator and the other proposed structure is made of K1 = K3 = 3.9, K2 = K4 = 25 in the top and bottom layer of gate insulator which are compared to the conventional low-k CNTFET with K = 3.9 for the whole sections of gate insulator in both layers. Fig. 5(a) depicts the transfer characteristics (I_D-V_{GS}) of the conventional low-k CNTFET (K=3.9) and two type of FDGI-CNTFET which include dielectric constant K1=K3=25, K2=K4=3.9 and also K1=K3=3.9, K2=K4=25 in the top and bottom layer of gate insulator are simulated and compared to each other. It is observed from Fig. 5(a) that the device with K1= K3 = 25 (high dielectric) near the source, K2 = K4 = 3.9 (low dielectric) near the drain in both layers and also K1 = K3 = 3.9 near the source, K2 = K4 = 25 (high dielectric) near the drain in both layers of gate insulator have a lower I_{OFF} and as well as higher I_{ON} in comparison with the conventional low-k CNTFET with K = 3.9, however the increment in I_{ON} is observed as very low. This is due to lower thermionic emission current near the barrier height in the middle of channel which led to less increase of I_{ON}. Moreover, the decrease in I_{OFF} is much significant and decrease in transmission probability is observed in Fig. 5 (c) and Fig. 5 (d). Therefore, the I_{ON} / I_{OFF} ratio has been a considerable decrease in comparison with the previous structures. Hence, they are not proper structures for CNTFET. Hence, less increase of I_{ON} and less growth in the number of propagating states cannot improve effectively the transmission probability in both proposed structures. Fig. 5(b) depicts the conduction band profile for the conventional low-k CNTFET (K = 3.9), and FDGI-CNTFET with dielectric constant K1 = K3 = 25, K2 = K4 = 3.9 and also with K1 = K3 = 3.9, K2 = K4 = 25 in the top and bottom layers of gate insulator. It is also observed that higher potential barrier near the drain for FDGI-CNTFET with K1= K3 = 25, K2 = K4 = 3.9 in comparison with the conventional CNTFET with K = 3.9 in both layers. In fact, both of proposed FDGI-CNTFET has a non-uniform barrier in the channel due to different dielectric constants at the drain and source sides of gate oxides. Their barriers are almost symmetrically toward each other in both layers. Hence, as shown in Fig. 5 (a), both layers behave opposite of each other in two proposed structures.

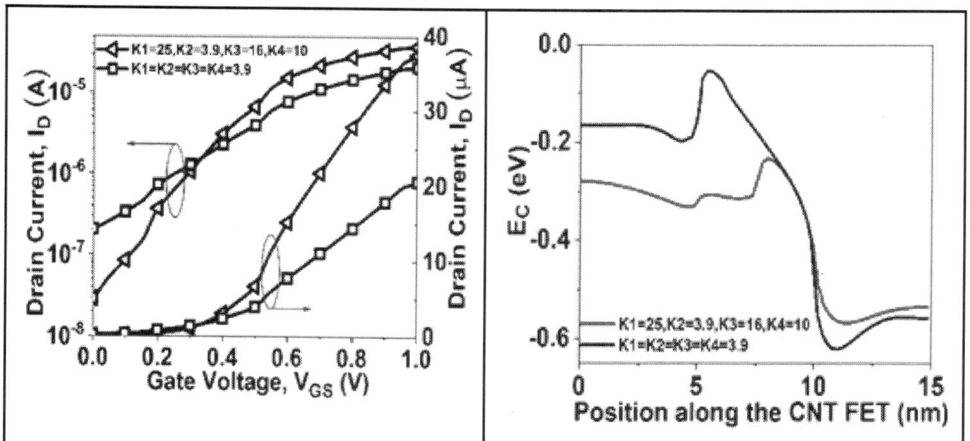

(a)	(b)

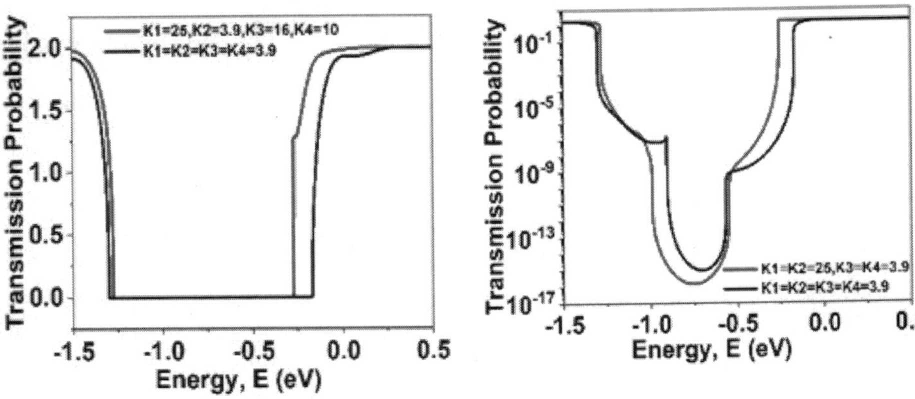

(c) (d)

Fig. 5. (a) Transfer characteristics of CNT FET at V_{DS} =0.5 V (b) CB profile of the CNT FET at ON state (c) Variation of transmission probability with energy in linear-scale (d) Variation of transmission probability with energy in log-scale.

CONCLUSION

In this work, a different structure of CNTFET with four different gate insulator is introduced combining the advantages of both high and low dielectrics. In the FDGI-CNTFET, to achieve a more efficient transistor, the gate length is divided into two equal sections in each layer (totally four). Simulations of the electronic properties of the proposed FDGI-CNTFET and the conventional CNTFET are based on solving the Poisson-Schrodinger equation self-consistently by means of non- Equilibrium Green Function (NEGF) and in the real space approach. In four sections, four different types of FDGI-CNTFET based on changing four different dielectric constants in gate insulator are studied and compared to CNTFET. Selecting a material with low dielectric constant for the gate insulator at the drain side can decrease the OFF-current more and also a material with high dielectric constant for the gate insulator at the source side can improve ON-current and reduce leakage current. Regarding to the results, the proposed FDGI-CNTFET compared to the conventional device, the maximum I_{ON}/ I_{OFF} ratio is achieved for FDGI-CNTFET with K1 = K2 = 25 and K3 = K4 = 3.9. Moreover, for FDGI-CNTFET with K1 = 25, K2 = 3.9, K3 = 16, K4 = 10, the I_{ON}/ I_{OFF} ratio is larger than conventional CNTFET. Finally, we suggest to investigate electrostatics of the proposed FDGI-CNTFET with higher I_{ON}/ I_{OFF} ratio analytically and experimentally for future work.

REFERENCES

[1] Skotnicki T., Hutchby J. A., King T. J., Wong H. S. P., and Boeuf F. 2005. The end of CMOS scaling: toward the introduction of new materials and structural changes to improve MOSFET performance. IEEE Circuits and Dev. Magaz., 21, 16–26.

[2] Rabaey J. M. and Malik S. 2008. Challenges and solutions for late-and post silicon design. IEEE Design and Test of Comp., 25, 296–302.

[3] Usmani F. A. and Hasan M. 2010. Carbon nanotube field effect transistors for high performance analog applications: an optimum design approach. Microelect. Journal, 41 395–402.

[4] Banerjee S., Chaudhuri A. and Chakrabarty K. 2020. Analysis of the Impact of Process Variations and Manufacturing Defects on the Performance of Carbon-Nanotube FETs. IEEE Transactions on Very Large Scale Integration (VLSI) Systems, 28, 1513-1526.

[5] Tamersit K., Madan J., Kouzou A., Pandey R., Kennel R. and Abdelrahem M. 2022. Role of Junctionless Mode in Improving the Photosensitivity of Sub-10 nm Carbon Nanotube/Nanoribbon Field-Effect Phototransistors: Quantum Simulation, Performance Assessment, and Comparison. Nanomaterials, 12, 1639.

[6] McEuen P.L., Fuhrer M.S. and Park H. 2002. Single-walled carbon nanotube electronics. IEEE Trans. Nanotechnol., 1, 78-85.

[7] Du¨rkop T., Getty S.A., Cobas E. and Fuhrer M.S. 2004. Extraordinary mobility in semiconducting carbon nanotubes. Nano Letters, 4, 35-39.

[8] Svensson J. and Campbell E.B. 2011. Schottky barriers in carbon nanotube- metal contacts. J. Appl. Phys., 110, 11-16.

[9] Appenzeller J. 2005. Comparing carbon nanotube transistors-the ideal choice: a novel tunneling device design. IEEE Trans. On Electron Devices, 52, 2568-2576.

[10] Roy K., Mukhopadhyay S. and Mahmoodi-Meimand H. 2003. Leakage current mechanisms and leakage reduction techniques in deep-submicrometer CMOS circuits. Proc. of the IEEE 91, 305-307.

[11] Sharma V., Pattanaik M. and Raj B. 2014. PVT variations aware low leakage INDEP approach for nanoscale CMOS Circuits. Microelectron. Reliab., 54, 90-99.

[12] Sharma V.K., Pattanaik M. And Raj B. 2015. INDEP approach for leakage reduction in nanoscale CMOS circuits. Int. J. Electron., 102, 200-215.

[13] Singh A., Khosla M. and Raj B. 2016. Comparative Analysis of Carbon Nanotube Field Effect Transistor and Nanowire Transistor for Low Power Circuit Design. J. Nanoelectron. and Optoelectron., 11, 388–393.

[14] Djamil R., Aicha K., Cherifa A. and Djeffal F. 2016. Impacts of high-k gate dielectrics and low temperature on the performance of nanoscale CNTFETs. J. Comput Electron, 15, 1-8.

[15] Information on http://vides.nanotcad.com/vides.

[16] Datta S. 2000. Nanoscale device modeling: The Green's function method. Superlattices Microstructure, 28, 253-278.

[17] Fallah M., Faez R. and Jafari A.H. 2013. Simulation of carbon nanotube field-effect transistor with two different gate insulator. Scientia Iranica, 20, 2332-2340.

[18] Owlia H. and Keshavarzi P. 2014. Investigation of the novel attributes of a double-gate graphene nanoribbon FET with AlN high-κ dielectrics. Superlattices Microstructures, 75, 613-620.

Chapter 12

Macromolecular Crowding and its Impact on Biomolecular Processes and in the Fields of Nanomaterials and Nanotechnology

Majumdar, Bibhab Bandhu[1]
[1]*VIT-AP university, Inavolu,*
Amaravati 522237, Andhra Pradesh, India
E-mail (bibhabbandhu.m@vitap.ac.in)

ABSTRACT

Intracellular environments consist of high concentrations of various macromolecules, such as proteins, nucleic acids, ions, metabolites etc. Together they occupy significant fractions: ~ 30% -40 % of entire cell volume. Macromolecular crowding In such heterogeneous crowded environments have been shown to impact the thermodynamics and equilibrium of various biomolecular processes such as, protein folding, biomolecular recognition, protein-protein interactions and self-assembly of biomolecules,polymers, nanoparticles etc. In this article, we discuss the recent challenges and developments in the field of macro molecular crowding. We show that the excluded volume interactions and other non-specific intermolecular interactions (hydrogen bonding, electrostatic interactions, hydrophobic interactions) need to be considered together for explaining the impact of crowding in vivo. We also discuss the impact of crowding on various biomolecular processes. Furthermore, we review some of the recent studies showing the use of crowding phenomena in the field of nano materials and nano technology.

INTRODUCTION

The intracellular environment area complex milieu of densely packed biomolecules, metabolites, ions etc. present at very high concentrations. [Ellis (2001)] They consist of proteins, nucleic acids, metabolites like:

phosphates, amino acids, carbohydrates, alcohols and ions, occupying 30%-40% of cytoplasmic volume. [Feig et al. (2017)] It is widely accepted that macromolecular crowding can have significant impact on structure and dynamics of various biomolecules, water soluble polymers and modify the equilibrium and thermodynamics of various bio molecular processes,such as: protein folding, protein-ligand binding, signal transduction, bio molecular diffusion, hydration,self-assembly of proteins etc. occurring in vitro and in vivo. [Gnutt et al. (2016), Feig and Sugita (2012), Harada et al. (2012)] Recently materials scientists employed macromolecular crowding to control self-assembly of molecules for tuning various physical properties and function of nanomaterials. [Hata et al. (2018)] In this review we summarizevarious aspects of macromolecular crowding and its effect on different biomolecular processes and molecular self-assembly and nanomaterials. We first discuss the basic aspects of crowding: the intermolecular interaction forces responsible for crowding effect and the available experimental and computational techniques to study biomolecular processes in crowded environments. We also review some of the recent studies focusing on the importance ofmacromolecularcrowdingphenomenainthe fieldofnanomaterialsandnanotechnology.

IMPACT OF CROWDING ON BIOMOLECULAR PROCESSES

Macromolecular crowding effect is primarily referred to as the consequence of repulsive excluded volume interactions between mutually impenetrable macromolecules. [Ellis and Minton (2003)] Macromolecular crowding phenomena have been described by the scaled particle theory of hard sphere fluids. [Reiss et al. (1959), Zhou et al. (2008)] Repulsive excluded volume interactions are shown to entropically stabilize more compact or folded configurations of a biomolecule over unfolded or extended conformations by reducing the volume excluded towards other molecules. Macromolecular crowding is shown to facilitate host-guest complex formation or protein-ligand binding process and oligomerization, protein aggregation,molecular self-assemblyetc. [Zhou et al. (2008), Majumdar and Mondal (2022), Minton (2000), Rivas and Minton (2016), Hartl et al. (2011), Hata et al. (2018)]Earlier experimental studies employed artificial crowders in vitro, such as: polyethylene glycol (PEG), Ficoll, Dextran, polyvinylpyrrolidone (PVP) etc. to mimic the intracellular environments. Hydration structure and dynamics of biomolecules are also affected in presence of macromolecular crowding. The self-diffusion coefficient and dielectric properties of solvents get significantly reduced in presence of concentrated protein crowder solutions. [Harada et al. 2012, Ando and Skolnick (2010)] The altered hydrodynamics in effect can significantly modify biomolecular diffusion in crowded conditions. The recent advancements of sophisticated experimental technologies like: in-cell NMR [Li et al. (2010)],in vivo FRET, fluorescence spectroscopy opened new avenues for studying biomolecular processes occurring in vivo in realistic biological environments. [Gnutt and Ebbinghaus (2016), Gnutt et al. (2015)]Computer simulation techniques, such asatomistic rigid body Brownian dynamics [Mc Guffee and Elcock (2010)], Monte Carlo [Prytkova et al. (2016)]and molecular dynamics (MD) simulation methods have emerged as promising tools for complementing experimental results. [Yu et al. (2016), Feig et al. (2017)] Recent advances of computational architectures like Anton supercomputers (ableper form 20 microseconds of full atomistic MD simulations of ~ million atoms in few hours) [Shaw et al. (2021)] and MD simulation techniques like enhanced sampling simulations, umbrella sampling, metadynamics, steered MD, enabled researchers to study biomolecular processes in crowded environments with improved spatial and temporal accuracy at a significantly reduced computational expense. Recent in vivo experimental studies and computer simulation studies reported counteracting enthalpic contributions to the overall crowding effect originating from various components of non specific interactions, such as: hydrogen bonding, electrostatic, hydrophobic interactions, showing significant deviations from predictions based on excluded volume effect. [Gnutt and Ebbinghaus (2016)]This clearly explains a need for considering excluded volume effect in combination with other non specific attractive interactions for explaining crow ding related phenomena.

RECENT APPLICATIONS OF MACROMOLECULAR CROWDING IN NANOTECHNOLOGY

Researchers in the field of materials science and nanotechnology have been using macro-molecular crowders to control the structure and function of various nanomaterials. [Hata et al. (2018)] Fmoc-dipeptides, an important precursor for making hydrogels and nanovehicles, have been used for gelation in presence of artificial crowders like PEG, resulting in enhanced rigidity of hydrogels. [Hassan et al. (2017)] Excluded volume effect provides additional stability to the assembly of supramolecular fibers. Macromolecular crowding has been employed to prevent diffusion-controlled aggregation of cellulose oligomers and for controlled self-assembly of cellulose oligomersinto nanoribbon-shaped crystals to form hydrogels. [Hata et al. (2017)] Self-assembly of nanomaterials like: gold nanorods, DNA functionalized gold nanoparticles (AuNPs), DNA wrapped carbon nanotubes (CNTs), get facilitated in presence of crowding. [Goodrich et al. (2004), Hata et al. (2018)] The impact of crowding on molecular self-assembly can also be described by the depletion interaction mechanism. The attractive depletion interaction is referred to as an attractive interaction force acting between two larger sized molecules and responsible for preferential depletion or exclusion of a smaller solute from the interface of the larger solutes. However, the preferential exclusion of the smaller solute or depletant are referred to as the repulsive depletion interactions between the smaller and the larger solutes. The basis of stabilization of self-assembly of nanoparticles is described as a consequence of attractive depletion interactions. In a contrasting scenario, colloidal dispersion of nanoparticles is also shown to get stabilized inpresence of crowding due to a repulsive depletion interaction between the nanoparticles,such as: AuNPs, graphene oxides, quantum dots etc. [Li et al. (2021)] Crowding has also been employed tostudy the impact on biomolecular applications of nanomaterials. AuNPs in conjugation with glutathione (AuGSH) has been studied as a nanoparticle based renal clearable. [Lira et al. (2018)] The interactions of AuGSH and proteins can be enhanced in presence of crowding when compared to dilute buffer solutions. Single chain nanoparticles (SCNPs) have been used as a model system to accurately predict the conformational behavior of intrinsically disordered proteins (IDPs)in presence of crowding. [Moreno et al. (2016)] Macromolecular crowding has also been shown to enhance the formation of DNA nanoparticles resulting from depletion interaction between adjacent DNA coilsinduced by PEG crowders. [Hou et al. (2016)] Nanoparticles have also been used to enhance crowding effects. For example: melting temperature of a 12-mer DNA can be enhanced in conjunction of AuNPs and high concentration of PEG.[Zaki et al. (2012)] Cellulose nanocrystals, nanoporous silica gels have been used to mimic macromolecular crowding effect on controlled self-assembly of nanostructures. [Hata et al. (2018)] These recent findings show exciting prospects for studying the self-assembly of nanoparticles and the generation of functional nanomaterials in presence of crowding for various applications like: drug discovery, generation of biomaterials, gene therapy etc.

SUMMARY

In this review we have summarized various aspects of macromolecular crowding. We have discussed the intermolecular interactions responsible for crowding effect. Further more, we have summarized some of the recent findings that show case the use of macromolecular crowders for preparation of self-assembled functional nanomaterials, the role of nanoparticles for mimicking crowded environments etc. The impact of crowding has been explained by acombination of various intermolecular forces, such as excluded volume interactions, depletion forces, nonspecific attractive interactions etc. A significant research effort is required toobtain a universal understanding of crowding effects applicable in different scenarios. This will enhance applicability of crowding phenomena in research areas, such as: biophysical chemistry, nanotechnology and materials sciences.

REFERENCES

[1] Ellis, R. J., 2001. Macromolecular crowding: Obvious but underappreciated. Trends Biochem.Sci.,26: 597-604.

[2] Feig, M.; Yu, I., Wang, P.-h., Nawrocki, G. and Sugita, Y. 2017. Crowding in Cellular Environments at an Atomistic Level from Computer Simulations. J. Phys. Chem. B, 121: 8009-8025.

[3] Gnutt, D. and Ebbinghaus, S. 2016. The macromolecular crowding effect - From in vitro into thecell.Biol.Chem.,397:37-44.

[4] Feig, M. and Sugita, Y. 2012. Variable Interactions between Protein Crowders and Biomolecular Solutes Are Important in Understanding Cellular Crowding. J. Phys. Chem. B,116:599-605.

[5] Harada, R., Sugita, Y., and Feig, M. 2012. Protein crowding affects hydration structure anddynamics.J.Am.Chem.Soc., 134: 48424849.

[6] Hata, Y., Sawada, T. and Serizawa, T. 2018. Macromolecular crowding for materials-directedcontrolledself-assembly.J.Mater.Chem.B,6:6344-6359.

[7] Ellis,R.J. and Minton,A.P. 2003. Cellbiology:Jointhecrowd. Nature, 425:27-28.

[8] Zhou,H.-X.,Rivas,G. andMinton,A.P. 2008.MacromolecularCrowdingandConfinement:Bio-chemical, Biophysical, and Potential Physiological Consequences. Annu. Rev. Biophys.37: 375-397.

[9] Majumdar, B. B.; Mondal, J. 2012. Impact of Inert Crowders on Host Guest RecognitionProcess.J.Phys.Chem.B, 126: 4200-4215.

[10] Minton,A.P. 2000.Proteinfolding: Thickeningthebroth.Curr.Biol., 10: R97-R99.

[11] Rivas, G. and Minton, A. P. 2016. Macromolecular Crowding In Vitro, In Vivo, and In Between.TrendsBiochem.Sci., 41: 970-981.

[12] Hartl, F. U., Andreas, B. and Hayer-Hartl, M. 2011. Molecular chaperones in protein folding and proteostasis. Nature, 475: 324-332.

[13] Ando, T. and Skolnick, J. 2010. Crowding and hydrodynamic interactions likely dominate in vivo macromolecular motion. Proc. Natl. Acad. Sci. U. S. A., 107: 18457-18462.

[14] Li,C., Wang,G.-F.,Wang,Y.,Creager-Allen,R.,Lutz,E.a.,Scronce,H.,Slade,K. M.,Ruf, R. a. S., Mehl, R. a. and Pielak, G. J. 2010. Protein ^{19}F NMR in Escherichia coli. J. Am.Chem.Soc.,132:321-327.

[15] Gnutt,D.,Gao,M.,Brylski,O.,Heyden,M. andEbbinghaus,S.Excluded volume effects in livingcells.Angew.Chem.Int.Ed.2015,54,25482551.

[16] McGuffee, S. R. and Elcock, A. H. 2010. Diffusion, crowding & protein stability in a dynamicmolecularmodelofthebacterialcytoplasm.PLoSComput.Biol.,6: e1000694.

[17] Prytkova, V., Heyden, M., Khago, D., Freites, J. A., Butts, C. T., Martin, R. W. andTobias, D. J. 2016. Multi-Conformation Monte Carlo: A Method for Introducing Flexibilityin Efficient Simulations of Many-Protein Systems. J. Phys. Chem. B, 120: 8115-8126.

[18] Yu, I., Mori, T., Ando, T., Harada, R., Jung, J., Sugita, Y. and Feig, M. 2016. Biomolecularinteractions modulate macromolecular structure and dynamics in atomistic model of abacterialcytoplasm. eLife,5: e19274.

[19] Shaw, D. E. et al. 2021. Anton 3: twenty microseconds of moleculardynamics simulation before lunch. Int. Conf. High Perform. Comput. Netw. StorageAnal.,pp.1-11.

[20] Hassan,M.M.,Martin,A.D. andThordarson,P. 2015.Macromolecular crowding and hydrophobic effects on Fmoc-diphenylalanine hydrogel formation in PEG: water mixtures. J.Mater.Chem.B,3: 9269-9276.

[21] Hata, Y., Kojima, T., Koizumi, T., Okura, H., Sakai, T., Sawada, T. and Serizawa, T. 2017. Enzymatic synthesis of cellulose oligomer hydrogels composed of crystalline nanoribbonnetworks under macromolecular crowding conditions. ACSMacro Lett., 6: 165-170.

[22] Goodrich, G. P., Helfrich, M. R., Overberg, J. J. and Keating, C. D. 2004. Effect of macromolecular crowding on DNA: Au nanoparticle bioconjugate assembly. Langmuir, 20:10246-10251.

[23] Li, X., Liu, X and Liu, X. 2021. Self-assembly of colloidal inorganic nanocrystals: nanoscale forces, emergent properties and applications. Chem. Soc. Rev. 50: 2074-2101.

[24] Lira,A.L.,Ferreira,R.S.,Torquato,R.J.,Zhao,H.,Oliva,M.L.V.,Hassan,S.A.,

[25] Schuck,P. andSousa,A.A. 2018. Binding kinetics of ultrasmall gold nanoparticles with proteins. Nanoscale, 10:3235-3244.

[26] Moreno, A. J., Lo Verso, F., Arbe, A., Pomposo, J. A. and Colmenero, J. 2016. Concentrated solutionsofsingle-chainnanoparticles:Asimplemod elforintrinsicallydisorderedproteinsundercrowdingconditions.J.Phys. Chem.Lett.,7:838-844.

[27] Hou, S., Trochimczyk, P., Sun, L., Wisniewska, A., Kalwarczyk, T., Zhang, X., Wielgus-Kutrowska, B., Bzowska, A., Holyst, R. 2016. How can macromolecular crowding inhibit biological reactions? The enhanced formation of DNA nanoparticles.Sci. Rep., 6: 1-11.

[28] Zaki, A., Dave, N. and Liu, J. 2012. Amplifying the macromolecular crowding effect usingnanoparticles.J.Am.Chem.Soc., 134: 35-38.

Chapter 13

Enhanced Performance of Lead-Free $MASnI_3$ Based Perovskite Solar Cells: Numerical Approach

Bhumika. E[1], N. Hima Varsha[1], Megha Unni[2], Sudagar Jothi[2*], Chandu DS[1#]
[1]*School of Electronics Engineering, VIT-AP University, Amaravati, INDIA*
[2] *School of Advanced Sciences, VIT-AP University, Amaravati, INDIA*
*E-mail: *sudagar.jothi@vitap.ac.in; #chandu.ds@vitap.ac.in*

INTRODUCTION

In the recent years, there has been tremendous progress in the development of perovskite solar cells (PSC) with power conversion efficiency (PCE) increasing rapidly from 3% since its inception in 2009 to over more than 25% till date. With this high PCE, it is certain that the perovskite solar cell with its single junction architecture can highly challenge the existing and commercial solar panels belonging to the first generation. However, the progress in the commercialization and the industrial market entry of PSCs is not straight forward owing to the serious factors that affect the solar cell long-term stability due to exposure to external environmental conditions like moisture, humidity, air, heat and temperature. In addition, the commonly used PSCs consists of toxic lead (Pb) material $MAPbI_3$, $FAPbI_3$ or $CsFAMAPbI_3$ which may be cause serious damage to the users when it comes in contact with air, or the material oxidizes and hydrates due to moisture. Therefore, these heavy metals are a hinderance to the shelf-life of the perovskite solar cells and also dangerous from the environmental perspective. Accordingly, more than achieving high PCE researchers are currently more concerned about the longevity of lead-based perovskite solar cells.

On the other hand, the hole transporting material (HTL) is an important functional layer in the perovskite solar cell. HTL plays a prominent role in blocking the electrons from the perovskite and helps in efficient extraction and transport of positively charged holes from the active layer to the top-electrode. In addition, an ideal HTL also separates the perovskite layer from the positively charged electrode and aids in improving the device stability of the perovskite solar cell by isolating the moisture, preventing metal corrosion due to electrodes, providing excellent energy level matching with the perovskite, thereby improving the open circuit voltage (V_{OC}). However, the most widely used HTL spiro-OMeTAD has poor conductivity and requires doping for enhancing the device performance. Especially, the most common dopant LiTFSI is highly hygroscopic and although high power conversion efficiencies (PCEs) are achieved using doped spiro-OMeTAD, the performance of the perovskite solar cell deteriorates rapidly due to the moisture intake.

The existing research shows that there is a need for design and fabrication of lead-free perovskite solar cells which are based on HTLs that (i) do not require additional doping, and (ii) do not affect the stability of the PSC. In this work, we present the design and numerical simulation of PSC based on n-i-p architecture with CuI as the hole transporting material (HTM). The performance of the device is also compared against the PSC based on spiro- OMeTAD HTM. The results illustrate that PSCs with CuI as HTM exhibits the highest efficiency of 22.1% and it is proposed that CuI is a suitable inorganic HTM for design and fabrication of lead-free PSCs.

MATERIALS AND METHODS

The methodology must be clearly stated and described in sufficient detail or with sufficient references. The simulation has been done by using an open-source numerical simulator to investigate the dimensions and electro-optical properties to achieve maximum power conversion efficiency for lead-free perovskite solar cells. This software works by utilizing the Poisson's equation and the continuity equation of both charge carriers. The solution of the differential equations required for the calculation of optimum dimensions are given as follows: Initially, the Poisson's equation is considered for the semiconductor as shown in Equation 1

$$\nabla^2\psi = q \times (n\ \ p + N_A - N_D) / \varepsilon \qquad (1)$$

Where, N_A is the concentration of the acceptor

N_D is the concentration of the donor

ψ is the electrostatic potential.

Now, the continuity equation is written for the semiconductor using Eq.2

$$\nabla.J_n\ q\ (\partial n/\partial t) = +qR \qquad (2)$$

$$\nabla.J_p + q\ (\partial p/\partial t) =\ qR \qquad (3)$$

Where, Jn is the electron current density

JP is the hole current density current density JP is the hole current density

R is the rate of carrier recombination.

With conduct current flow in semiconductor there would be two effects. Primarily, the influence of electric field will affect the drift of minority charge carrier concentration and secondly, the concentration gradient is used to evaluate the diffusion current in the semiconductor. The relation between the drift-Diffusion current is given by the continuity equations:

$$J_n = q_n\mu_n E + {}_qD_n\nabla_n \quad (4)$$

$$J_p = q_p\mu_p E - {}_qD_p\nabla_p \quad (5)$$

Where, D_p represents hole diffusion coefficient and

Dn represents electron diffusion coefficient

Copper iodide (CuI), like most binary elements based on metal halide is an inorganic polymer. CuI has inherent advantages of higher bandgap, low cost of production, excellent chemical and morphological stability, and superior hole mobility. On the other hand, the organic material spiro-OMeTAD (2,2',7,7'-tetrakis(N,N-di-*p*-methoxyphenylamine)9,9'- spirobifluorene) is highly expensive for fabrication and it adds to the production cost of PSCs.

RESULTS AND DISCUSSION

The findings and arguments of the work should be explicitly described and illustrated. Supporting figures, tables and images of the The structure of the PSC used in thiswork follows the n-i-p architectureandconsistsofaFTO(Fluorine-dopedtinoxide)substrate,Titaniumdioxide(TiO2)astheelectron,MASnI3astheactivelead-freeperovskitelayer.Twodifferentholetransportinglayers (CuI/Spiro-OMeTAD) are used in this study. The back contact is made of gold (Au) asshownin Figure 1. The electro-optical properties of the materials used in the PSC areshown in Table1. Duringsimulation,thefollowingpropertieshavebeenvaried:

(i) The thickness of the perovskite, the TiO2, along with the hole transport layers

(ii) The corresponding mobilities of ETL and HTL

(iii) The doping concentrations of the electron and hole transporting materials

iv) The material bandgap.

In general, after the AM1.5 spectrum illumination, the perovskiteactive layer absorbs the energy and generates electron–hole pairs. These charge carriers due to the excited energy are then diffused to there spective electrodes via perovskite-HTL, perovskite-ETL interfaces and reach therespectiveelectrodes. Especially, this rate of hole extraction and transfer is more superior in case of CuI as HTM because of it's excellent matching of the HOMO energy level with MASnI3.As a result, it canbe seen that the FF of the cell with CuI as HTM (82.16%) is higher than that of the cell with spiro-OMeTAD as HTM (63.78%) as shown in Table 2.

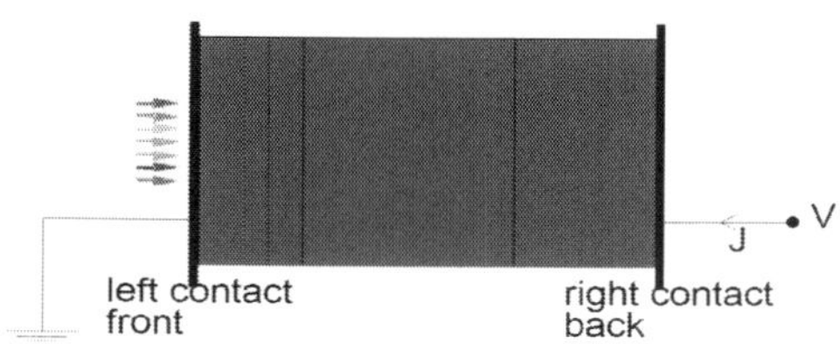

Figure1. $MASnI_3$ perovskite solar cell architecture.

This is the reason for the enhanced efficiency of the PSC with CuI (22.1%) compared to spiro-OMeTAD (16.84%). There is a significant improvement in the PCE by about 30% using CuI as HTM in lead-free PSCs. The photovoltaic (PV) parameters of the proposed cell structure are strongly influenced by the thickness of the MASnI3 layer. In this research, the optimum value of the perovskite layer thickness is 400 nm which is obtained by tuning the thickness range from 250 to 1000 nm. The obtained J-V characteristics and the extracted output parameters are summarized and showed in Figure.2(a). Table.2 shows the parameters PCE, FF, VOC and JSC reported for PSCs based on spiro-OMeTAD and CuI hole transporting materials.

Table1. S. Solar cell parameters and their values used in this study

Solar cell parameters	FTO	TiO_2	$MASnI_3$	CuI	Spiro OMeTAD
Material thickness (nm)	100	50	330	Variable	Variable
Band-gap(eV)	3.6	3.2	1.3	3.4	2.88
Electonaffinity(eV)	4.	4	4.17	2.1	2.05
Dielectric permittivity	9	9	6.5	10	3
CB effective density of states (cm^{-3})	2.2 10^{18}	2.0 10^{18}	1.0 10^{18}	2.5 10^{18}	2.5 10^{18}
VB effective density of states (cm^{-3})	1.8 10^{19}	1.8 10^{19}	1.0 10^{19}	1.8 10^{19}	1.8 10^{19}
Electron thermal velocity (cm/s)	1.0 10^{7}	1.0 10^{7}	1.0 10^{7}	1.0 10^{7}	1.0 10^{7}
Hole thermal velocity (cm/s)	1.0 10^{7}	1.0 10^{7}	1.0 10^{7}	1.0 10^{7}	1.0 10^{7}
Electronmobility (cm2 /V.s)	20	20	1.6	2.0 10^{4}	2.0 10^{-4}
Holemobility (cm^2/V.s)	10	10	1.6	2.0 10^{4}	2.0 10^{-4}
Shallow donor density N_D(cm^{-3})	1.0 10^{18}	5.0 10^{17}	—	—	—
ShallowAcceptordensity N_A(cm^{-3})	—	—	1 10^{16}	2.0 10^{19}	2.0 10^{19}
Defectdensity N_t(cm^{-3})	1 10^{15}	1 10^{15}	2 10^{15}	1 10^{15}	1 10^{15}

Table 2. PV parameters of the lead-free PSC with CuI and spiro-OMeTAD as HTL.

Performance parameters	CuI	Spiro-OMeTAD
Open voltagecircuit (V)	0.90	0.89
Shortcircuitcurrent density(mA/cm^2)	29.63	29.47
Fillfactor(%)	82.16	63.78
Efficiency (%)	22.1	16.84

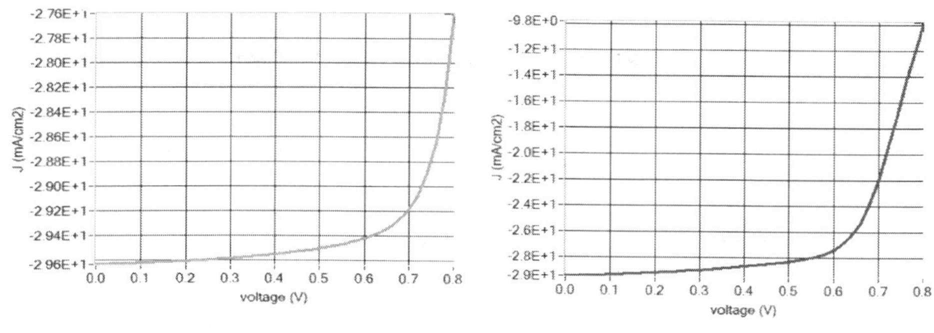

(a) (b)

Figure 2. J-V characteristic of (a) Spiro-OMeTAD, and (b) CuI as HTM

The cell was simulated to determine the influence of variation in thickness of CuI and Spiro- OMeTAD on open-circuit voltage (VOC), short-circuit current density (JSC), fill factor (FF) shown in Figure 2.(a) and Figure 2.(b). The thickness of CuI and spiro-OMeTAD are varied from 5 nm to 100 nm and the other layer parameters are kept constant. It is observed that the best performance in terms of PCE is observed for a thickness of 70 nm (in case of spiro- OMeTAD) and 45 nm (in case of CuI). Generally, it was found that PSCs with inorganic HTLs exhibited higher PCE than that with organic HTLs, according to the simulation the efficiencies attained by CuI is 22.1% to that of spiro-OMeTAD is 16.84% undertaking perovskite thickness 330nm. Figure 3.(a) and 3.(b) show the corresponding quantum efficiencies (QE) of the proposed lead-free PSCs with CuI and spiro-OMeTAD. The cell with CuI as HTM shows excellent photon absorption at the visible spectrum which is consistent with the FF and JSC values in the J-V characteristics. Note that the cell with CuI as HTL demonstrates a higher JSC in the 100-200 nm regions as compared to Spiro-OMeTAD. The improved performance of the perovskite solar cells with CuI as HTL layer (compared to Spiro-OMeTAD) is mainly attributed to the high hole mobility/ HTM conductivity of these copper-based p-type semiconductors.

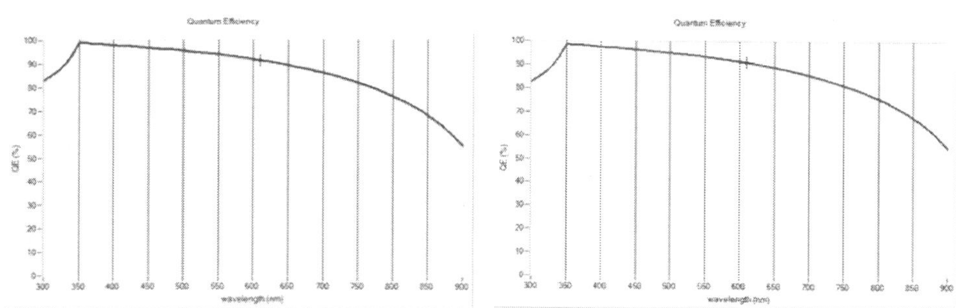

(a) (b)

Figure3. EQE spectra of the PSC of (a) spiro-OMeTAD, and (b) CuI

CONCLUSION

In this work, Spiro-OMeTAD being organic and Copper Iodide being inorganic HTLs were discussed and reviewed by numerical simulation on SCAPS-1D. It was found in our simulation that PSCs with inorganic HTL exhibited higher PCE than that with organic HTL, while adding dopants were indispensable for most of the effective organic HTL to improve their conductivity and/or hole mobility, inducing instability of high-performing spiro-OMeTAD. For CuI, the highly efficient PSCs with it could reach a PCE over 20%, and more importantly, they showed superior stability, making inorganic HTL great potential. CuI and Spiro-OMeTAD thicknesses were varied to investigate their influence on the device performance. The simulation revealed that the CuI thickness between 40nm to 5nm and Spiro-OMeTAD between 40nm to 60nm is suitable for better device performance. This clearly shows that lead-free perovskite solar cells at high-performance are feasible to achieve, by enhancing the device parameters, in future research.

REFERENCES

[1] Anand Kumar Singh, Arup Mahapatra, Basudev Pradhan, Jayanta Kumar Baral, Shristy Srivastava. 2021. Performance optimization of lead free-MASnI3 based solar cell with 27% efficiency by numerical simulation. 1-3.

[2] Song Li, Wen-Hua Li, Yong-Li Cao, Zhi-Shan Bo.A. 2021. A brief review of hole transporting materials commonly used in perovskite solar cells. 2720-2721.

[3] Anand Kumar Singh, Basudev Pradhan, Prashant Kumar, Shristy Srivastava. 2021. Comparative Performance Analysis of Lead-Free Perovskites Solar Cells by Numerical Simulation. 12-15.

[4] Minemoto T, Murata M. 2014. Device modeling of perovskite solar cells based on structural similarity with thin film inorganic semiconductor solar cells. J Appl Phys 116:054505.

[5] Wei X et al. 2017. Numerical simulation and experimental validation of inverted planar perovskite solar cells based on NiOx hole transport layer. Superlattices Microstruct 112:383– 393 17.

Chapter 14

Orientation in A Large-Area Conducting Polymer Thin-Films for Organic Field-Effect Transistors Fabricated by Ribbon-Shaped FTM

Atul S.M. Tripathi*

*School of Electronics Engineering, VIT-AP University Amravati, A.P.-522237, India*E-E-mail: atul.tripathi@vitap.ac.in*

ABSTRACT

In the recent past, organic semiconductors emerged as a new class of low-cost semiconducting materials for the fabrication of flexible organic electronics devices. Having unique solution processable properties makes it for large area fabrication and easily deposited on a desired substrate. Including the solution processable, extended -conjugation in the main chain favors the charge transport.

Currently, a number of deposition techniques are under practice to control the thin film morphology. Among some are the following molecular ordering approach to align the thin film to enhance the device's efficiency, but they still have some issues in terms of multilayer deposition, mechanical and compositional damage.

Here we focused on a recently developed thin film deposition method named ribbon- shaped floating film transfer method (FTM) for the fabrication of thin films, which avoids charge accumulation, mechanical damage and solubility under-layer along with the least material wastage. In this study, ribbon shaped FTM was utilized to fabricate a large area of thin films of organic semiconductors and their optoelectronics characterizations.

For optimizing the thin film, follow the controlling parameters such as liquid substrate, casting temperature and concentration (w/w) of semiconductor materials. One optimized condition, like EG:GL-3:1 viscosity of liquid substrate, 1% (w/w) concentration of semiconducting material and 50 0C as a casting temperature.

It has been observed that this method provides large- area oriented thin films (L=20 cm x W= 2 cm) shape. The films show optical anisotropy in terms of Dichroic ratio (DR) from 2 to 20. Further, oriented thin film is utilized as an active layer along the channel direction during the OFETs fabrication and shows high field effect mobility in the order of 10^{-2} to 10^{-3} Cm2/Vs. Moreover, it is also showing electrical anisotropy and reflecting from the optical anisotropy.

INTRODUCTION

Over the Over the last decade there has been a tremendous development in optoelectronic devices such as Organic field effect transistors (OFETs) [1], electronic paper, liquid crystal displays (LCDs), organic light emitting diodes (OLEDs) [2] and solar cells [3] and photodetector [4]. Amongst organic electronic devices, OFETs bear the fascination for research and development, being an integral part of many devices owing to their amplifying and switching actions. [5] OFETs performance depends on the quality of thin film morphology. However, the cost depends upon the thin-film deposition technique. Currently, the existing number of methods for deposition of thin film, including some proposed molecular orientation techniques. However, most of them use shear forces resulting in face-on orientation. At the same time, the existing possibility of physicochemical damages to the underlying soft polymeric layers also cannot be ruled out.[6] However, direct deposition of thin films via slow solvent evaporation from the polymer solution inherently results in thermodynamically favoured edge-on orientation of Conjugated polymers CPs, which is highly favourable for planar devices like OFETs.[7] To resolve such issues for the thin film fabrication of solution processable CPs, our group have developed a technique for the facile fabrication of thin films on any desired substrate by casting a polymer solution on an orthogonal liquid substrate named as dynamic floating film transfer method (FTM) [8]. It not only provides multilayer thin films with minimal material wastage but also imparts the orientation in the fabricated thin films. Conventional FTM leads to thin films in a circular shape due to the spread of the films in all directions and films on the outer ends are relatively thin and more oriented. To resolve this issue, we further improvised this technique using an assisting slider to provide directionality during the spreading of the polymer solution, leading to a large area of ribbon-shaped thin films named as ribbon shaped FTM. [9-10] In this article, we report the low cost ease of fabrication technique that shows optical as well as electrical anisotropy. The prepared thin film in the opposite spreading direction (parallel orientation) provides improved optoelectronics performance by more than one order as compared to the spin coating method.

MATERIALS AND METHODS

In this work, Ethylene Glycol and glycerol were utilized as viscous liquid substrate. For making the solution, various polymers were utilized after dissolving in dehydrated chloroform solvent. Among the CPs, a few are lab synthesized and some are purchased from the sigma Aldrich. Respective CP solutions with concentrations of 1% and 0.5% (w/w) were prepared for thin film fabrication by FTM and spin coating, respectively. Large area oriented thin films were prepared with ribbon shaped FTM. For comparison, non-oriented spin coated thin films were also prepared by spinning the chloroform solution of respective CPs at 3000 rpm for 120 s. The oriented films prepared by ribbon shaped FTM were transferred onto the substrates like glass and SiO2 grown silicon wafer for optical and electrical characterizations, respectively. Polarized electronic absorption spectra were measured by a UV–visible spectrophotometer (JASCO V-570) equipped with a Glan-Thompson prism. For electrical characterization, OFETs were fabricated using highly doped p-type Si substrate with a 300 nm thermally grown SiO2 insulating layer having capacitance (Ci =10 nf/cm2) SiO2 surface was treated by octadecyltrichlorosilane (OTS) for the formation of hydrophobic self-assemble monolayer, which assists the better adhesion of the CPs during thin film fabrication on the SiO2 insulator. Oriented floating solid films were transferred onto OTS treated SiO2/Si substrates by stamping. OFETs were fabricated in a bottom gate top contact (BGTC) device architecture, where 50 nm of source and drain gold electrodes were deposited at the rate of 1.5 Å/s by a thermal evaporator with high vacuum of 10-6 torr using nickel shadow mask. In order to make parallel and perpendicular OFETs for the investigation of anisotropic charge transport, direction of the mask was accordingly selected (considering direction of orientation in the film) to deposit top source-drain electrodes. The channel length (L) and width (W) were fixed for all of the devices to be 20 μm and 2 mm, respectively. Electrical characterizations were done by computer controlled 2 channel electrometer (Kiethley 2612) at vacuum of 10 -3 Pa.

RESULTS AND DISCUSSION

In organic electronics, device performance depends upon many factors, but controlling the thin film morphology is one of the most important parameters. Currently, a number of thin film deposition techniques (spin coating, dip coating and spray coating) have been used to deposit the thin film but they provide isotropic film with a resultant low device performance. However, some methods (mechanical rubbing, friction transfer method, and solution flow and strain alignment) provide the anisotropic film to enhance the device's performance. Although these methods are favorable for orienting the thin film, there are still some drawbacks such as mechanical, compositional damage and bottleneck for multilayer casting.

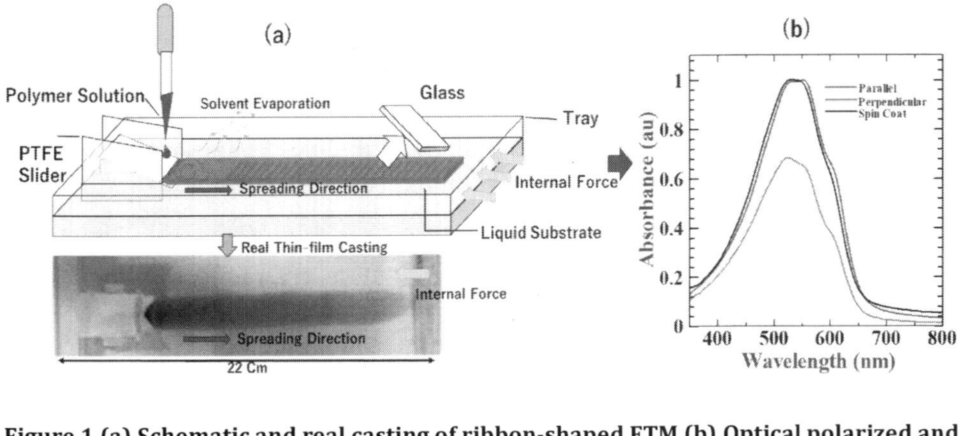

Figure 1 (a) Schematic and real casting of ribbon-shaped FTM (b) Optical polarized and non-polarized absorbance of alkyl thiophene based conjugated polymer.

Here, the ribbon-shaped FTM resolves the existing problem and provides the oriented thin film. Figure 1 (a) shows the schematic and real film casting of a ribbon-shaped FTM. The prepared solid film polarized absorption (parallel and perpendicular) is shown in figure 1 (b) and compared with the spin cast method. The optical absorption range is 400 nm to 700 nm and shows 552 nm of absorbance in parallel FTM, while 524 nm, 538 nm in perpendicular and spin cast respectively. The film prepared in parallel shows a red shift of 14 nm with a more clear vibronic shoulder as compared to the perpendicular and spin cast. The optical absorbance is showing less in perpendicular, meaning polymer chain alignment along the parallel direction is expressed in terms of dichroic ration (DR) given in equation 1 and its value depends upon controlling the parameter and polymeric backbone structure.

$$DR = \frac{Maximum\ Absorption_{\parallel}\ at\ (\lambda max_{\parallel})}{Absorption_{\perp}\ at\ (\lambda max_{\parallel})} \quad \ldots\ldots\ldots\ldots\ldots\ldots\ldots\ (1)$$

The oriented thin is transferred into the channel across the source and drain electrode. Device IV characteristics are measured in parallel and perpendicular directions by shifting the position of the shadow mask. Figure 2 (a) shows the device structure and its output and transfer characteristics plotted in 2 (b), (c) respectively. Device parameters are measured from transfer characteristics following drain current equation (2) at the saturation region. There are remarkable enhancements in device mobility over the spin cast method.

$$I_{DS} = \frac{W}{2L}\mu C_i (V_{GS} - V_{TH})^2 \ \ldots\ldots\ldots\ldots\ldots\ (2)$$

Moreover, it is also showing charge transport anisotropy and correlated with optical anisotropy. Apart from this, other parameters like On/Off ratio and threshold voltage were improved.

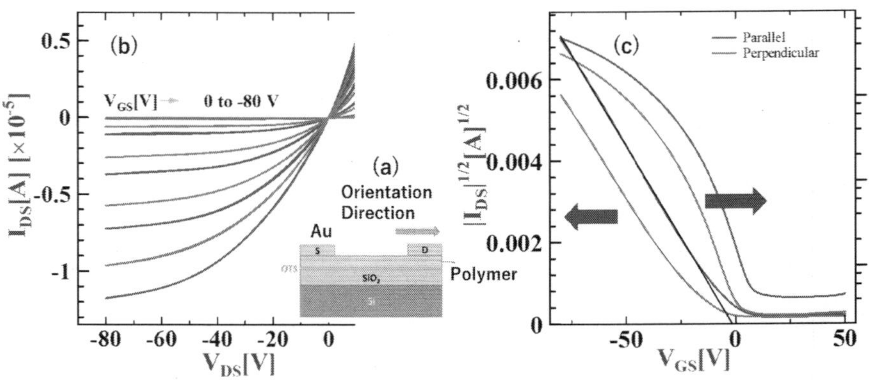

Figure 2 (a) Bottom gate top contact device structure (b) output characteristics (c) transfer characteristics.

It is interesting to note that this method provides oriented thin film with utilizing least material usage and ease of fabrication. This method is adequate for various CPs . [11] Moreover, it is also suitable for molecules and composite materials. [12] FTM prepared thin film provides the edge-on-orientation over the other orientation technique and that is good for OFETs. The FTM coated device shows high charge carrier transport even at a weaker orientation. [13] Here, it is clear that in figure 1 (b) showing a very weak orientation (DR ~ 1.5) still the charge carrier mobility is 2 10-2 cm2/Vs.

CONCLUSION

In this report, one novel floating thin film fabrication technique is applied to fabricate a large area ribbon-shaped up to the size of a cm scale (20 cm in length and 2 cm in width) by dropping 15 µl of polymer solution. This thin film deposition technique not only minimizes the wastage of material and cost but also provides an edge on oriented thin film which is suitable for planer devices like OFETs. The parallel oriented thin film showing redshift over the perpendicular and spin coating and showing high charge carrier transport up to 10-3 to 10-1 Cm2/Vs.

After applying this approach, we achieved DR up to 1.4 - 20 showing high optical anisotropy. Further, this result is reflected in the device by utilizing the FTM coated thin film across the source to drain direction and calculate the electrical anisotropy. The resultant optical and electrical anisotropy reflect on each other and are proved from the in-plane GIXD measurement in an earlier publication.

ACKNOWLEDGMENT

The authors thanks to the Prof. Shyam S Pandey Kyushu Institute of Technology, Japan, for the support.

REFERENCES

[1] C.D. Dimitrakopoulos, P.R.L Malenfant, Organic Thin Film Transistors for Large Area Electronics. Adv. Mater. 14 (2002) 99 117.

[2] H. Uoyama, K. Goushi, K. Shizu, H. Nomura, C. Adachi, Highly Efficient Organic Light-Emitting Diodes from Delayed Fluorescence. Nature 492 (2012) 234 238.

[3] Huynh, W. U.; Dittmer, J. J.; Alivisatos, A. P. Hybrid Nanorod- Polymer Solar Cells. Science 2002, 295, 2425 2427.

[4] D. S Chang, H. Kong, H. Effect of Solvent on Detectivity of Solution-Processed Polymer Photodetectors. Opt. Lett. 38 (2013) 2814 2817.

[5] R. Dost Ray, S. K. Das, A. Grell, M. Oscillator Circuit Based on a Single Organic Transistor Appl. Phys. Lett. 93 (2008) 113505.

[6] M. Brinkmann, L. Hartmann, L. Biniek, K. Tremel, and N. Kayunkid, Macromol, Rapid Commun. 35 (2014) 9

[7] M. Pandey, A. Gowda, S. Nagamatsu, S. Kumar, W. Takashima, S. Hayase, and S. S. Pandey, Adv. Mater. Interfaces 428 (2018)1700875.

[8] T. Morita, V. Singh, S. Nagamatsu, S. Oku, W. Takashima, K. Kaneto, Enhancement of Transport Characteristics in Poly(3- Hexylthiophene) Films Deposited with Floating Film Transfer Method. Appl. Phys. Express 2 (2009) 111205.

[9] A.S.M. Tripathi, M. Pandey, S. Sadakata, S. Nagamatsu, W. Takashima, S. Hayase, S.S. Pandey, Anisotropic charge transport in highly oriented films of semiconducting polymer prepared by ribbon-shaped floating film, Appl. Phys. Lett. 112 (2018), 123301,

[10] A.S.M. Tripathi, N. Kumari, S. Nagamatsu, S. Hayase, S.S. Pandey, Facile fabrication of large area oriented conjugated polymer films by ribbon-shaped FTM and its implication on anisotropic charge transport, Org. Electron. Phys. Mater. Appl. 65 (2019) 1–7

[11] A.S.M. Tripathi, N. Kumari, S. Nagamatsu, S. Hayase, S.S. Pandey, Facile fabrication of large area oriented conjugated polymer films by ribbon-shaped FTM and its implication on anisotropic charge transport, Org. Electron. Phys., Mater. Appl 65 (2019) 1–7,

[12] A.S.M. Tripathi, K. Kondratenko, B. Duponchel , N. Hurduc , I. Carlescu b Y. Boussoualem , A. Daoudi Thin Solid Films 742 (2022) 139044.

[13] M. Pandey, S. Nagamatsu, S.S. Pandey, S. Hayase, W. Takashima, Enhancement of carrier mobility along with anisotropic transport in non-regiocontrolled poly (3- hexylthiophene) films processed by floating film transfer method, Org. Electron. 38 (2016) 115–120.

Chapter 15

Review on Perovskite Solar Cells: Materials and Stability

Vissam Setti Mourya Teja[1,] Chinmaya Kumar Swain[1,] Arun Kumar[2] and L. Sowjanya Pali[2]

[1]VIT-AP University, Andhra Pradesh, 52237, India.
[2]J.C. Bose University of Science and Technology, YMCA, Faridabad, 121006, India.
Correspondence: lakshmi.sowjanya@vitap.ac.in

ABSTRACT

Perovskite solar cells have attracted tremendous interest in the photovoltaic technology with power conversion efficiencies exceeding 26% outshining the well-established inorganic thin film solar cell technologies. With most work focused on methylammonium lead trihalide ($MAPbX_3$) perovskites, investigations on complete or partial replacement of cations (MA^+, Pb^{+2}) and anions (X^-) opened new pathways for favourable modifications in the band gap of perovskite materials for efficient light harvesting, charge carrier transport and stability. In this review, we briefly discussed the notable improvements in power conversion efficiencies of single junction solar cells fabricated using various perovskite materials and internal and external causes for instability in the perovskite solar cells. This review is helpful in gaining deep insight into the choice of materials for developing high performance, stable and low-cost perovskite solar cells.

INTRODUCTION

Increasing world energy demands and the scarcity of fossil fuels in the near future led to the search for alternative energy sources. Solar energy has emerged as one of the most promising alternative energy sources for the production of electricity over the last four decades [John S. Mangum (2021), Ling Hong (2020), L. Sowjanya Pali (2021)].

Over a decade, from the fabrication of the first solid-state single-junction perovskite solar cell with an efficiency of 9.7% in 2012 [Hui-Seon Kim (2012)] to 25.6%

efficiency in 2021 [Tianhao Wu (2021)], perovskite solar cell technology showed unprecedented advancement in achieving the high-power conversion efficiencies over the other photovoltaic technologies. Reported power conversion efficiencies of perovskite solar cells from 2012 to 2021 are shown in figure 1.

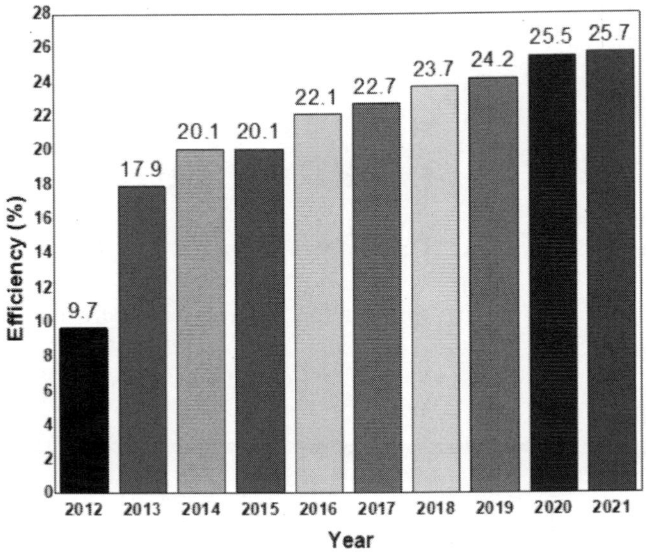

Fig 1: Reported power conversion efficiencies of perovskite solar cells (NREL Efficiency Tables)

Rapid progress in the power conversion efficiencies of these perovskite solar cells can be attributed to their unique features such as wide tunablility of bandgap, broad absorption spectrum, high carrier mobilities, long diffusion lengths and ease of fabrication on both rigid and flexible substrates of the perovskite materials [Stefaan De Wolf (2014), Samuel D. Stranks (2013), Guichuan Xing (2013)]. Further abundancy of these perovskite materials on earth and cost-effective solution processable fabrication methods make this perovskite photovoltaic technology superior to other photovoltaic technologies. While perovskites have excelled in power conversion efficiency (PCE), their advances in stability are significantly more complex and have progressed more slowly. The evolution of efficiency has always outpaced stability as rapid innovations in efficiency inherently require longer duration validation to ensure that they do not introduce new degradation pathways; this is especially true for emerging technologies. In the paper, we discussed the power conversion efficiencies and causes of structural instability of perovskite materials in solar cells.

PEROVSKITE MATERIALS

Perovskite materials are described by ABX_3 structure shown in figure 2, where A is a large monovalent cation, B is a much smaller divalent cation, and X is an anion that bonds to both the A and B cations.

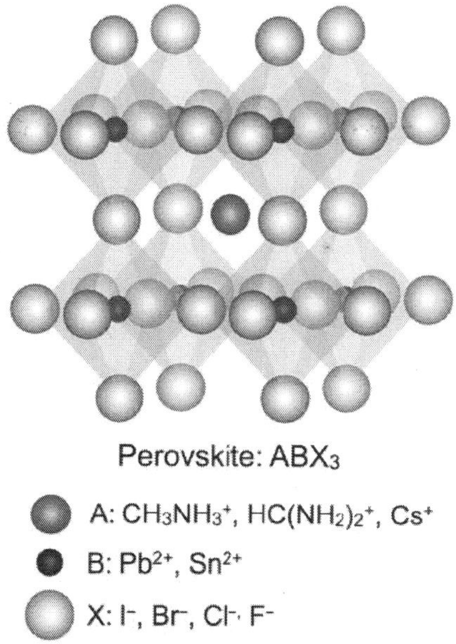

Fig 2: Crystalline structure of Perovskites [Brittman, Sarah (2015)]

The ability of elements A, B, and X to form stable perovskite structure can be predicted by the Goldschmidt tolerance factor, t. The Goldschmidt tolerance factor is calculated from the ionic radius of the atoms.

$$t = \frac{R_A + R_X}{\sqrt{2}(R_B + R_X)}$$

Where R_A, R_B and R_X are the ionic radii of corresponding atoms. When 't' is less than 0.8 perovskite materials shows orthorhombic structure, when 't' from 0.8 to 1 these materials are having cubic structure and when 't' is above 1 perovskite materials will have layered structures.

Among perovskite materials, Organometal halide perovskites are the main class of photovoltaic material for the fabrication of solar cell in which A is an organic cation such as methylammonium ($CH_3NH_3^+$) and formamidinium ($NH_2CH=NH_2^+$), M is a divalent metal cation Cu^{2+}, Ni^{2+}, Co^{2+}, Fe^{2+}, Mn^{2+}, Pd^{2+}, Cd^{2+}, Ge^{2+}, Sn^{2+}, Pb^{2+}, Eu^{2+}, and X is a monovalent anion namely a halogen F , Cl , Br , I , etc.

With a band gap of 1.55 eV, high electron and hole mobilities, $MAPbI_3$ is most studied perovskite material for the fabrication of solar cell. Recent studies showed 21% efficiency with $MAPbI_3$ as light absorbing layer. However, $MAPbI_3$ suffers phase transition at approximately 56 C forming distorted tetragonal perovskite structure at higher temperatures. This phase transition causes detrimental effect on the charge carrier transport, thus on the performance of solar cell. Further, the MA cations also absorbs moisture from air resulting in the faster degradation of solar cell. Replacing methylammonium ion (MA^+) with formamidinium ion ($NH_2CH=NH_2^+$) showed stable perovskite structure till 130 C. Mixed perovskites $FAMAPbI_3$ showed a high absorbance in the infrared region from 780 to 900 nm. It is essential to replace toxic lead in order to expand the use of these perovskites. To create perovskite compounds with enhanced stabilities and improved efficiency, it is preferable to substitute materials in the device design or combine cations, anions, and halides.

(a) Replacement of MA with Other Molecules

With 1.48 eV, Formamidinium lead iodide ($FAPbI_3$) shows broad absorption spectra than methylammonium lead iodide ($MAPbI_3$). Figure 3 shows experimental bandgap values of different perovskite materials. $FAPbI_3$ showed a high absorbance in the infrared region from 780 to 900 nm. The inclusion of formamidinium cation results in shifting the incident photon-to-current efficiencies (IPCE) onset towards a longer wavelength and is shown in figure 4.

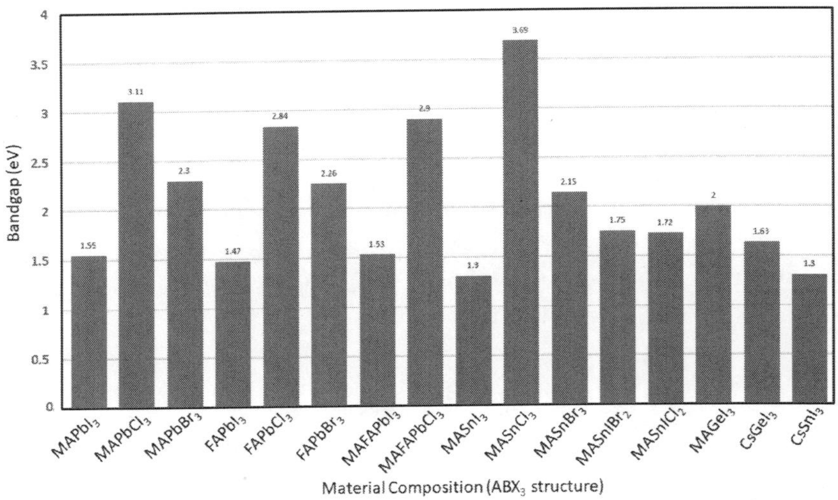

Fig 3. Experimental band gap values of different compositions of perovskite material [Hao Feng (2014), Hao Feng (2014) Ali Roshan (2018)]

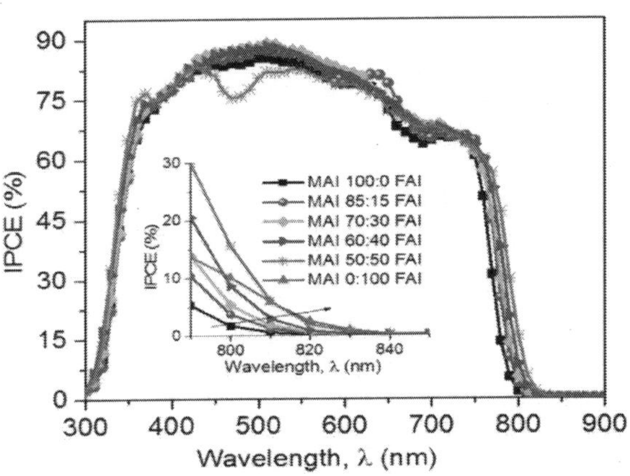

Fig 4: IPCE measurements of perovskite solar cell with different MA and FA concentrations [Salado Manuel 2016].

Mixed perovskites $FAMAPbI_3$ showed a high absorbance in the infrared region from 780 to 900 nm. Recently, $FAPbI_3$-based PSCs attained a validated efficiency of 25.21%. According to these findings, some of the $FAPbI_3$ and other mixed halide perovskites exhibit good potential for effective PSCs which are given in the below table 1.

Table 1: Efficiencies of different perovskite absorbing devices

Material Combinations	Efficiency (%)	References
$FAPbI_3$	20.10	Yang,Woon Seok, (2015)
$FAPbI_3$	24.82	Mingyu Jeong, (2020)
$FAPbI_3$	25.21	Jaeki Jeong (2021)
$FASnI_3$	8.30	Jokar Efat (2019)
$MASnI_3$	6.40	Noel Nakita K (2014)
$MASnI_3$	7.78	Pengcheng Wang (2020)
$FASnI_3$	7.72	Liu Gengling (2020)
$FAMAPbI_2$	21.25	Zhengchun Wang (2022)
$CsSn_{0.5}Ge_{0.5}I_3$	7.11	Min Chen (2019)
$Rb_5Cs_{10}FAPbI_3$	20.35	Turren-Cruz (2018)
$Cs_x(MA_{0.17}FA_{0.83})_{(100-x)}Pb(I_{0.83}Br_{0.17})_3$	21.10	Saliba, Michael (2016)

It was found that the addition of the formamidinium cation in the MAI slows down the crystal formation in perovskites and allows the formation of large crystals and a high degree of order. Inorganic cation tuning, using Rb and Cs, enables highly crystalline formamidinium-based perovskites without Br or MA. Here with a perovskite design of $Rb_5Cs_{10}FAPbI_3$, Silver-Hamill Turren-Cruzand co-workers achieved a stabilized PCE of 20.35% [Turren-Cruz (2018)]. Saliba, Michael et al., also used a mixture of a triple Cs/MA/FA cation, to achieve high-efficiency perovskite solar cells with a stabilized PCE at 21.1% [Saliba, Michael (2016)]. moreover, Cesium chloride (CsCl) is added to the double cation (Cs, formamidinium) perovskite precursor to produce the extremely phase-stable perovskite without the use of MA. The bromide (Br) amount is also controlled due to the bromide's (Br) inadequate bandgap, which results in high-power conversion efficiency. It can be seen that without MA PSCs with appropriate content of CsCl and $PbBr_2$ simultaneously can reach the best-integrated performance.

(b) Lead-free Perovskites

Lead (Pb) toxicity is a barrier to the widespread commercial manufacture of lead halide perovskites and their use in solar cells. Literature suggests first row of transition metals, which include Mn^{2+}, Ni^{2+}, Cu^{2+}, Zn^{2+}, Ag, Sn, and Bi are best suited replaceable elements for Pb. Lead can be replaced in a tetragonal or cubic perovskite unit cell with multiple alkaline earth metal cations like Be, Sr, Ca, Mg, etc of oxidation state +2 while maintaining crystal symmetry. Additionally, compared to $MAPbI_3$, $MAGeI_3$ exhibits better hole and electron conductivity and stability. For mono-replacements, a few lead-free alternative perovskites ($MASnI_3$ and $MAGeI_3$) can be successfully synthesized A mixed study of the transition elements shows that MA (Pb: B) I_3 (B = Co, Cu, Fe, Mg, Mn, Ni, Sn, Sr, and Zn), with the highest efficiency of 17.2% for the molar ratio of 63Pb:1Co. For MA(Ca/Si) Br_3 and MA(Ca/Si) Cl_3, the calculated PBE band gap values are, 2.26, 1.25, 1.52, 1.73, 2.86, and 2.09 eV, while for $MAZn_{0.5}Si_{0.5}Br_3$, and $MAZn_{0.5}Si_{0.5}Cl_3$ the calculated PBE band gaps are 1.73 and 2.2 eV, respectively. It should be observed that $MACa_{0.125}Si_{0.875}I_3$ has a band gap of 1.54 eV by HSE calculations, which is very close to the experimental value of $MAPbI_3$ (1.55 eV).[Ali Roshan (2018)]. Hence, it could be an ideal replacement for $MAPbI_3$ with high efficiency in PV cells. The ratio adjustment of Ca/Si and Zn/Si provides a way to improve the properties of the material, such as photovoltaic performance and electronic transport properties.

Replacing Pb with Sn

The solid-state device based on the $CH_3NH_3SnI_3$ perovskite shows a high mean short-circuit photo current density J, an open-circuit voltage, and a moderate fill factor corresponding to a PCE of 5.23%. In addition to the significant improvement in Voc, an increase in FF was also observed upon the incorporation of the Br ions. Amongst the investigated $CH_3NH_3SnI_3Br_x$ perovskites, the device with $CH_3NH_3SnIBr_2$ delivered the highest PCE of 5.73% [HaoFeng (2014)].

The easy oxidation of Sn and Ge is also considered to be the main problem for limiting the development of Sn-based and Ge-based PSCs. In order to enhance the film quality of perovskites and prevent Sn^{2+} oxidation, additional unique compound have also been used as additives [Kour Ravinder (2019)].

Replacing Pb with Ge

Germanium is another candidate for the substitution of lead for lead-free perovskite solar cells because of its valence electronic configuration as that of Pb^{2+}, Ge^{2+} has a small ionic radius. A study of Ge-based perovskite $AGeX_3$ reported the estimated values of optical bandgap for $CsGeI_3$ (1.63 eV), $MAGeI_3$ (2.0 eV), and $FAGeI_3$ (2.3 eV) [Kour,Ravinder (2019)].

The easy oxidation of Sn and Ge is also considered to be the main problem for limiting the development of Sn-based and Ge-based PSCs. In order to enhance the film quality of perovskites and prevent Sn^{2+} oxidation, additional unique compounds have also been used as additives.

(c) Replacement of Iodide in Perovskites

Methylammonium (MA) lead perovskites with mixed iodide bromide compositions, $MAPb(I_{1x}Br_x)_3$, are reported to exhibit improved stability. Recent research showed that using bromide instead of iodide to make mixed-halide perovskite films results in extended operational lives and excellent power conversion efficiency. The time is taken for the perovskite sample to degrade increases with increasing bromide content. The introduction of bromide ions into $MAGeI_3$ perovskites enhances not only photovoltaic performance but also stability to a slight extent. The substitution of 10% of the iodide content by bromide results in $MAGeI_{2.7}Br_{0.3}$ perovskite that reported an SPCE of 0.57% as a light absorber in solar cells. [Kopacic, I (2018)].

Researchers also observed that adding a small content of Br to the perovskite absorbing layer also makes it as a hole conducting layer. The solar cell consists of conductive glass and thin TiO_2 film that consists of perovskite composition according to the formula $CH_3NH_3PbInBr_{3n}$ (where 0 n 3). Hole conductivity has occurred through the hybrid $CH_3NH_3PbInBr_{3-n}$.

Further, the absorption spectra of the $MAPbBr_{1.5}I_{1.5}$ films with and without iodide treatment are presented in the figure 5 showing that the mixed halide films exhibit absorption below 700 nm with an exciton peak appearing at 630 nm. The I_3 treated films exhibited slightly higher absorbance than the untreated ones. It is interesting to note that the iodide treatment did not introduce any noticeable spectral shifts or changes in the overall spectral profile as both treated and untreated films exhibited excitonic peaks around 630 nm.

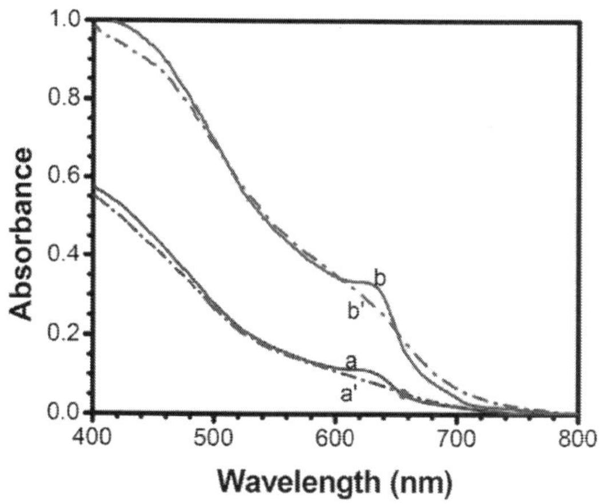

Fig 5: The absorption spectra of $MAPbBr_{1.5}I_{1.5}$ films (a,a') without and (b,b') with iodide treatment. The spectra were recorded (a,b) before and (a'b') after exposure of visible light for 30 minutes [Balakrishna, R.G. (2018)].

It is also observed that by intermolecular changes there is a change in bandgap and ionization energy as well as electron efficiency of the material used in the device describes that the adjustable perovskite structure enables a number of multi-factor ways to control the perovskite's properties. The features of perovskite could be efficiently customized by altering the concentrations of chemical composition. Perovskite is appropriate for a variety of optoelectrical applications due to its simple modification process.

STABILITY OF PEROVSKITE SOLAR CELLS

(a) Perovskite Intrinsic Stability

The organic cation has little impact on the electrical properties or band structure of the perovskite layer. Its main role is to balance charge within the crystal structure. However, the size of the "A-cation" can have a noticeable effect on lattice structure.

The subsequent size of the B-X bond can also affect the band gap and stability of a perovskite solar cell. Therefore, by varying the ratios of A-cations in the precursor, lattice qualities can be tuned. The suitability of an A-cation for use in a perovskite structure is determined by its tolerance factor (t).

Different cation sizes can result in lead halide octahedral distortion and tilting. If this tilting is significant, it can shift the phase of the perovskite crystal into an undesirable state for photovoltaic activity. This is represented by the tolerance factor. If the tolerance factor is 1, then this produces a cubic perovskite structure. However, if the tolerance value falls between 0.7 and 0.9, then the tetragonal or orthombic phases can be induced. $MAPbI_3$ is tetragonal at room temperature. Cubic, Tetragonal and Orthorhombic structures are shown in figure 6.

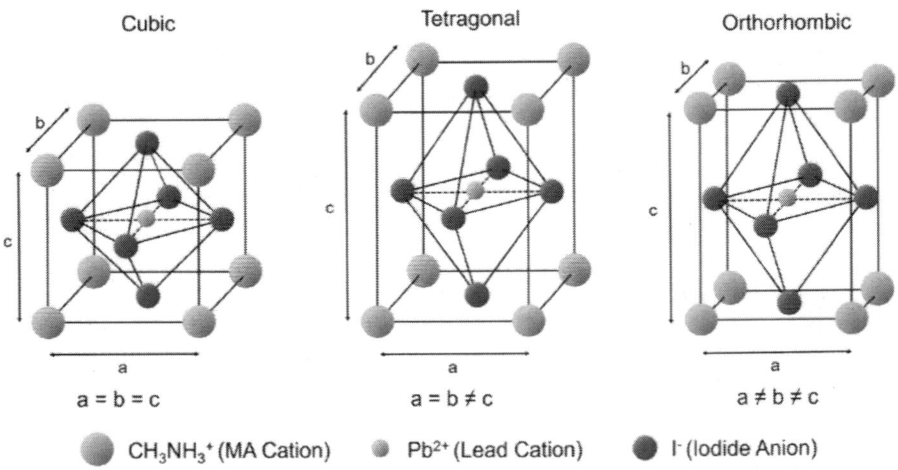

Fig 6: Crystal structures of perovskites (a) Cubic, (b) Tetragonal and (c) Orthorhombic [Thomson, S (2018)].

The A and B cations then provide one and two electrons, respectively, to charge balance three positive X anions. This forms a bandgap between the unoccupied B cation's p-orbital and occupied X anion's p-orbital. While the tolerance factor is a good 0th order approximation, it has been shown to mis predict 26% of structures tested. Due to this, Bartel et al. proposed a modified tolerance factor, τ. [Bartel CJ (2019)]

$$\tau = \frac{R_X}{R_B} - n_A\left(n_A - \frac{R_A/R_B}{\ln\left(R_A/R_B\right)}\right)$$

where R_A, R_B, and R_X have the same definitions, n_A is the oxidation state of the A cation, and $\tau < 4.18$ indicates perovskite formation with increasing probability. While the new modified tolerance factor does not predict alternate structures, it predicts perovskite structures much more accurately (8% incorrect) and provides a likelihood of formation. As a result, using a combination of the two, a functional estimate of structural stability can be made for neat compositions. Given the wide variety of formulations possible, we will limit our discussion to a subclass of perovskites that better resemble those studied in 2009 by Kojima et al. than the classical "perovskite" ($CaTiO_3$) discovered by Lev Perovski in 1839. This subclass of metal-halide perovskites has relevant photovoltaic properties such as a band gap in the visible light regime. To achieve this, they utilize tin (Sn) or lead (Pb) at the B-site and iodide (I), bromine (Br), and/or chlorine (Cl) at the X-site. Given the B- and X-site components, appropriate A-sites such as methylammonium (MA), formamidinium (FA), and cesium (Cs) can then be chosen to obtain suitable tolerance factor.

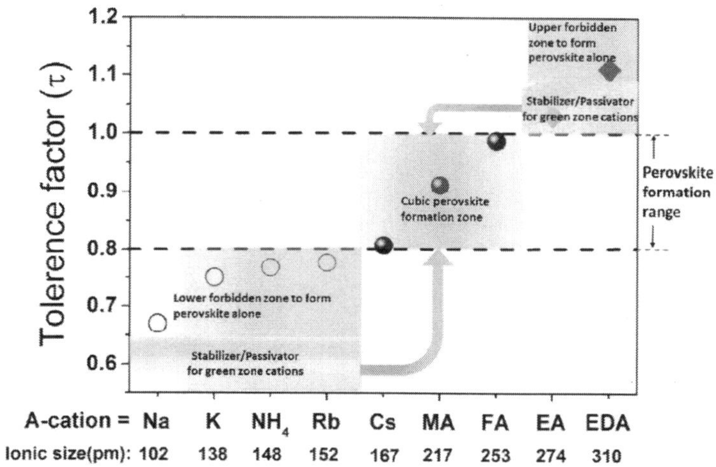

Fig 7: Calculated tolerence factor for different cation like MA,FA,Cs. [Han G, 2018].

Methylammonium lead halides undergo phase transitions from orthorhombic to tetragonal and then to cubic as temperature is increased whereas caesium lead halides undergo phase transitions from orthorhombic to cubic. These phase transition temperatures are highly dependent on composition. It is noteworthy that these changes can have a drastic effect on the optoelectronic properties of perovskites; many noncubic phases are photovoltaically inactive, such as room temperature structures for $FAPbI_3$ and $CsPbI_3$. As such, these neat compositions are considered "phase unstable" using this evaluation and would not be expected to make stable solar cells. However, it has been shown that alloys can be used to improve phase stability.

(b) Perovskite Extrinsic Stability

Despite perovskites general robustness to internal defects, external factors such as light, oxygen, temperature, and humidity (water) have been shown to interact with the perovskite material. These interactions can induce general phase changes by altering the structure's tolerance factor or induce chemical reactions that have the potential to cause irreversible degradation, as will be discussed in this section. Although the details of these mechanisms are not well known for most formulations, they have been studied most extensively for $MAPbI_3$ given its prevalence in literature. Moreover, due to its appropriate tolerance factor and single A-, B-, and X-sites, $MAPbI_3$ offers a phase pure system where chemical reactions, rather than phase transitions, dominate material degradation. This has allowed the community to develop an unconvoluted mechanistic understanding of chemical degradation in $MAPbI_3$ that can be used to predict how and why altering a given lattice site alters degradation.

i. Light (Photostability)

Like other semiconductors, $MAPbI_3$ forms an excited state when it absorbs light. In this state, it has been shown to have increased ionic mobility due to the creation of iodide vacancies. It has been theorized that this is due to the oxidation of iodide atoms by photogenerated holes. This transformation from iodide to iodine causes a reduction in ionic size, allowing it to leave the lattice and form an interstitial and vacancy. While it is unclear to what extent this reaction is reversible, the creation of charged iodide vacancies and free iodine could lower the activation energy for various degradation pathways, create new degradation mechanisms, and/or alter device behaviour.

ii. Oxygen (Photooxidative Stability)

As mentioned in the last subsection, light increases iodide vacancies and ionic mobility in the perovskite. Aristidou. Et. al., showed that in presence of light, oxygen intrusion results in the formation of superoxide (O_2^-) ion. When coupled with oxygen, this leads to irreversible degradation into PbI_2. [Aristidou. N (2017)]. Figure 8 shows the schematic of oxygen induced photo degradation. The absorbance profiles of $MAPBI_3$ films shown in figure 9. These results showed that the absorption profile is almost same after aging for 48 hrs in dry air and dark, where as in presence of light and oxygen there is a drastic decrease in the absorption of perovskite films. This indicates the strong influence of light and open air in reducing the power conversion efficiency of a solar cell.

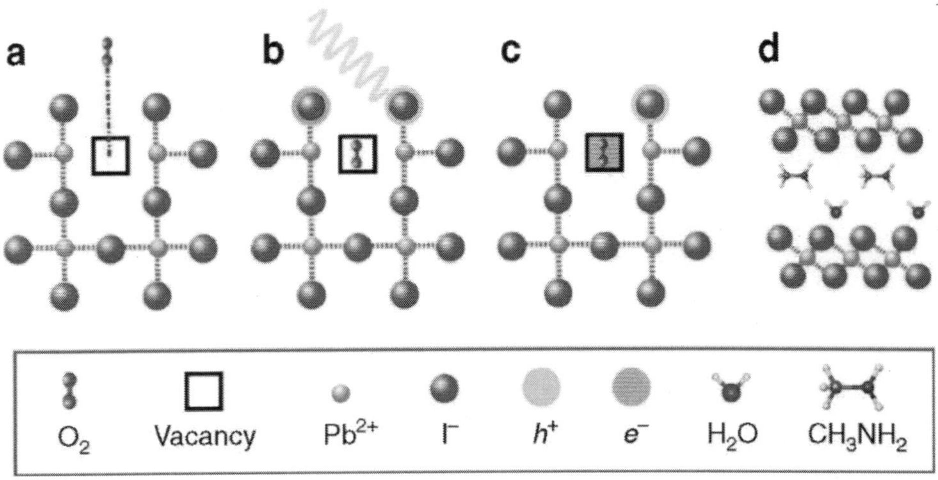

Fig 8. The reaction of O_2 with $CH_3NH_3PbI_3$. (a) Oxygen diffusion and its interaction with lattice, (b) electrons and holes creation by photoexcitation of $CH_3NH_3PbI_3$ (c) superoxide creation (d) reaction and degradation to layered PbI_2, H_2O, I_2 and CH_3NH_2. [Aristidou. N (2017)]

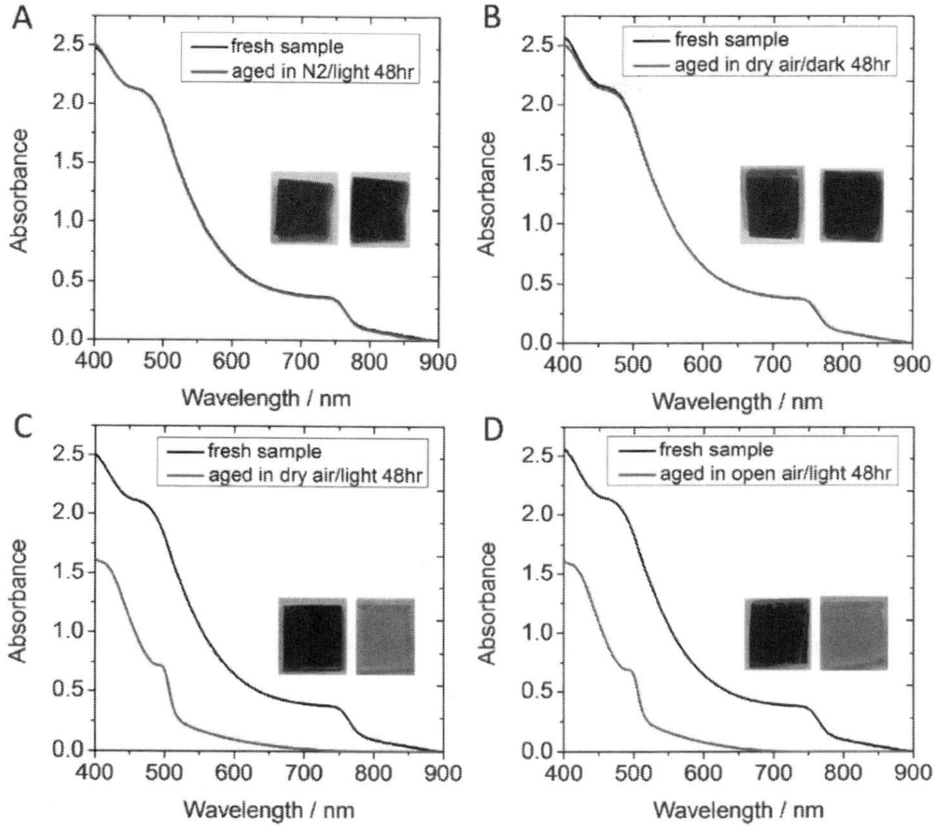

Fig 9: Absorbance profile of $MAPbI_3$ film before and after aging in various conditions (Inset shows the optical images of the films) [Bryant, D (2016)]

Haque et al. suggestedthat molecular oxygen reversibly adsorbs and diffuses through iodide vacancies, where they create a state near the conduction band [Aristidou, N (2019)]. If these states are present when the perovskite is excited, they trap a photoexcited electron, forming a highly reactive and charged O_2 superoxide. To become neutral, this oxide initiates an acid–base reaction with the MA cation, deprotonating it to form water, iodine, lead-iodide, and methylamine gas,

$$O_2(g) \xrightarrow{MAPbI_3^{+}} O_2^{\cdot}(g)$$

$$CH_3NH_3PbI_3(s)+O_2^{\cdot-}(g) \xrightarrow{\text{Deprotonation}} CH_3NH_2(g)+PbI_2(s)+\frac{1}{2}I_2(s)+H_2O(l)$$

Although the deprotonation process is afforded by the methylammonium cation, the general process of photooxidation, where an excited electron in the conduction band is oxidized by molecular oxygen, is common among most neat photovoltaic absorbers due to the prevalence of long-lived carriers in the excited state. However, when these materials are incorporated into a solar cell, the density of electrons in the conduction band is drastically reduced as they are quickly extracted by the electron transport layer (ETL). Consequently, the degree to which the perovskite oxidizes is highly dependent on the relative rates of oxidation and electron charge transfer to another material and will be much smaller for a perovskite incorporated into a device or on an ETL.

iii. Temperature (Thermochemical Stability)

In addition to oxygen and light, $MAPbI_3$is also sensitive to temperature. Degrading into PbI_2 in inert environments at temperatures as low as 85 C, as can be seen by the conductive atomic force microscopy (C-AFM) image in figure 10.

Fig 10: Conductive atomic force microscopy(C-AFM) image of pristine $MAPbI_3$ (left) films subjected to 85°C for 24 hours in N_2(right). Here percentage represent the fraction of area that contributes to current. [Conings, B (2015)]

Studies of the thermogravimetric analysis (TGA) agree that this occurs through an initial sublimation of the organic halide components, followed by decomposition of the metal halide at much higher temperatures.

$$CH_3\,NH_3\,PbI_3\,(s) \rightleftharpoons CH_3\,NH_2\,(g) + HI(g) + PbI_2\,(s)$$

$$CH_3\,NH_3\,PbI_3\,(s) \longrightarrow CH_3\,I(g) + NH_3\,(g) + PbI_2\,(s)$$

It is noteworthy that the products of the first reaction have been demonstrated to form perovskite when used as precursors, while the products in the second have been demonstrated to form nonprimary ammonium salts, denoted by the reversible and irreversible arrows. However, lamination studies using a combination of temperature and pressure show improvements in material properties when temperatures sufficient for thermal degradation are utilized and mass transport is prevented. As such, both reactions are likely reversible in systems where constituents are not prone to evaporate away or be consumed by other reactions [Dunfield, S. P(2018)].

iv. Water (Hygroscopic Stability)

Finally, water and other polar solvents/vapors have been demonstrated to affect the perovskite due to their ability to form solvated phases. Although degradation from polar solvents can be avoided by controlling processing environments and using orthogonal solvents to deposit sequential layers, completed devices will necessarily be exposed to humid environments when fielded. As such, significant effort has been made understanding interactions between water vapor and $MAPbI_3$. In general, it has been shown that water molecules initially hydrate the perovskite crystal structure, forming a monohydrate phase in single crystals.

$$4(CH_3NH_3)PbI_3+4H_2O \rightleftharpoons 4[CH_3NH_3PbI_3.H_2O]$$

$$4[CH_3NH_3PbI_3.H_2O] \rightleftharpoons (CH_3NH_3)_4PbI_6.2H_2O+3PbI_2+2H_2O$$

In both cases, these reactions are reversible. As such, mono and di hydrated perovskites stored in inert environments with no external input kinetically revert to their dehydrated phase with some minor limitations due to phase separation of the products. However, while incorporation of water does not initially cause permanent degradation, it does induce significant structural deformation by forcing the $[PbI_6]^4$ octahedra to separate. This separation causes the dehydrated structure to transform from a 3D network of octahedra to a 1D chain of octahedra for the monohydrate and a 0D framework of isolated octahedra for the dihydrate.

In addition to destroying the photovoltaic properties of the material, this structural deformation weakens the bonds between the organic cation and the $[PbI_6]^4$ octahedra, allowing for easier deprotonation of the cation and/or degradation of the film through external energy inputs such as heat and/or electrical bias This generally reversible hydration occurs until the perovskite structure has been saturated with water, at which point the organic component of the perovskite structure starts to dissolve, irreversibly creating lead iodide and aqueous methylammonium iodide. For the single crystal, this phase is suggested to evolve directly from the monohydrate phase [Askar, A. M(2017),Leguy, A.M (2015)]. For thin films this phase is evolved from dihydrate phase.

$$CH_3NH_3PbI_3.H_2O(s)+H_2O(l) \rightarrow CH_3NH_3I(aq)+PbI_2(s)+2H_2O(l)$$

$$(CH_3NH_3)_4PbI_6.2H_2O+H_2O(l) \rightarrow 4CH_3NH_3I(aq)+PbI_2(s)+2H_2O(l)$$

Crystal structure of $MAPbI_3$ at different hydrated states is shown in figure 11. Discrepancies between these reactions are likely explainable by the differences between the defect structure of thin films and single crystals. Specifically, thin films have grain boundaries or a higher surface-area-to-bulk ratio and therefore possess more defects with different formation energies and transition levels than their single crystal counterparts. These defects may make the formation of the dihydrate phase energetically favourable to the irreversible dissolution of the organic component, causing it to form first.

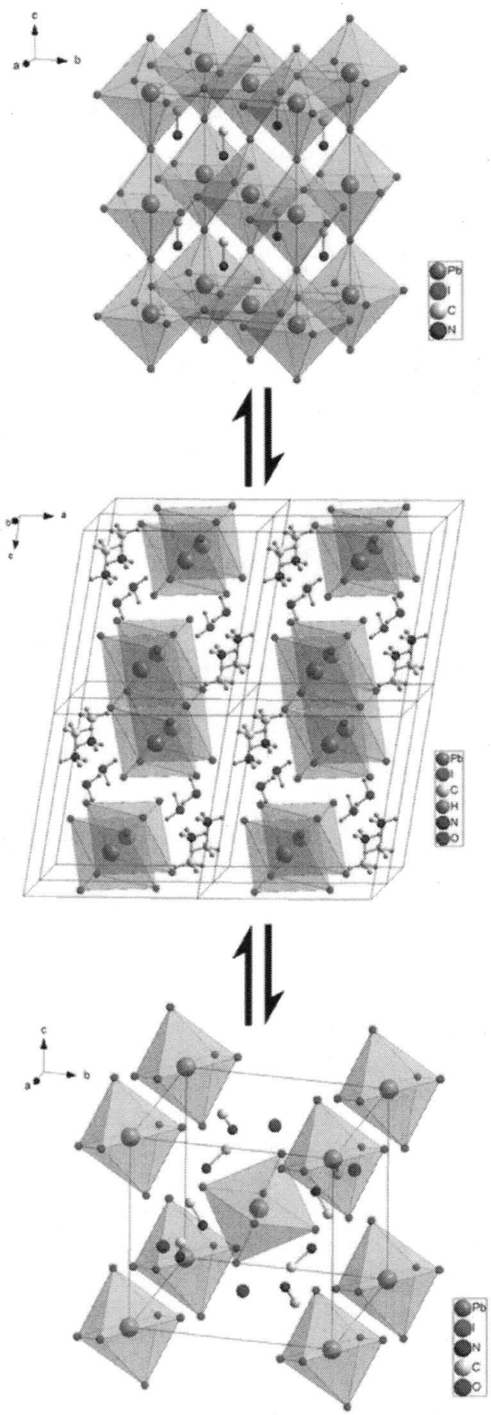

Fig 11: Structure of cubic $MAPbI_3$ (left), monohydrate $MAPbI_3$(middle), and dihydrate $(MA)_4Pb_62H_2O$ (right) [Leguy, A.M (2015)]

However, due to the intrinsic ionic characteristics of hybrid perovskite film, the stability of devices, especially in operating conditions is a critical concern in perovskite-based photovoltaics [Park, N. G (2016), Lee, J. W(2015), Nie, W(2016)]. It has become manifested that perovskite films generally begin to degrade from the defects at surfaces and grain boundaries, which are more susceptible to degradation when contact with water and oxygen [Chu, Zhaodong (2017), Yun, Jae S (2015), Yun, Jae Sung (2018)]. Therefore, minimizing the density of defective locations, such as grain boundaries areas and handling the surface dangling bonds are very important to achieve high performance and device stability. Various post-processing methods, such as pressure-induced crystallization, solvent annealing, chemical vapor recrystallization, Ostwald ripening and two-dimensional liquid cage annealing were usually used to repair crystal defects and reconstituted crystal structure with dramatic enhanced grain size [Yun, Jae Sung (2018), Do Hyung Chun (2019), Yang, Yuqian (2020), Yi, Ahra (2019), Long, Mingzhu (2017), Li. Wenzhe (2018)]. This conclusion is supported by the single crystalline PSC with dramatic improved stability comparing to their multi-crystalline counterparts. The introduction of Cs (1.81 Å) to fully replace MA can further minimize instability factors, such as the MA ion migration and $MAPbI_3$ phase decomposition in conditions of humidity, light or heat and etc [Lee, Jong Woo(2019), Correa-Baena(2017), Zhou, Huawei(2015), Juarez-Perez, Emilio J(2016)].

SUMMERY

Recent progress in the power conversion efficiencies of perovskite solar cells is mainly cantered at the organolead perovskites and their mixed compositions. With huge possibility of compositional variation, the structural flexibility organolead perovskites materials out-pass their inorganic counterparts in terms of power conversion efficiencies, cost-efficiency, high throughput and ease of device fabrication. Apart from bandgap tuning and phase stability associated with the compositional variation in organolead perovskites, improved film morphology and proper choice of electron and hole materials will further enhance the power conversion efficiency and stability of perovskite solar cell making them promising class of materials for future energy conversion applications.

REFERENCES

[1] Ali Roshan, Guo-Jiao Hou, Zhen-Gang Zhu, Qing-Bo Yan, Qing-Rong Zheng, and Gang Su. 2018. Predicted lead-free perovskites for solar cells. Chemistry of Materials 30, 3, 718-728.

[2] Aristidou, N., Eames, C., Sanchez-Molina, I., Bu, X., Kosco, J., Islam, M. S., and Haque, S. A. 2017. Fast oxygen diffusion and iodide defects mediate oxygen-induced degradation of perovskite solar cells. Nature communications, 8(1), 1-10.

[3] Askar, A. M., Bernard, G. M., Wiltshire, B., Shankar, K., and Michaelis, V. K. 2017. Multinuclear magnetic resonance tracking of hydro, thermal, and hydrothermal decomposition of $CH_3NH_3PbI_3$. The Journal of Physical Chemistry C, 121(2), 1013-1024.

[4] Balakrishna, R.G., Kobosko, S.M. and Kamat, P.V., 2018. Mixed halide perovskite solar cells. Consequence of iodide treatment on phase segregation recovery. ACS Energy Letters, 3(9), pp.2267-2272.

[5] Bartel, C.J., Sutton, C., Goldsmith, B.R., Ouyang, R., Musgrave, C.B., Ghiringhelli, L.M. and Scheffler, M., 2019. New tolerance factor to predict the stability of perovskite oxides and halides. Science advances, 5(2), p.eaav0693.

[6] Brittman Sarah, Gede Widia Pratama Adhyaksa, and Erik Christian Garnett. 2015. The expanding world of hybrid perovskites: Materials properties and emerging applications. MRS communications, 5, 1, 7-26.

[7] Bryant, D., Aristidou, N., Pont, S., Sanchez-Molina, I., Chotchunangatchaval, T., Wheeler, S., Durrant, J.R. and Haque, S.A., 2016. Light and oxygen induced degradation limits the operational stability of methylammonium lead triiodide perovskite solar cells. Energy & Environmental Science, 9(5), pp.1655-1660.

[8] Chu, Zhaodong, Mengjin Yang, Philip Schulz, Di Wu, Xin Ma, Edward Seifert, Liuyang Sun, Xiaoqin Li, Kai Zhu, and Keji Lai. 2017. Impact of grain boundaries on efficiency and stability of organic-inorganic trihalide perovskites. Nature communications 8, no. 1, 1-8.

[9] Conings, B., Drijkoningen, J., Gauquelin, N., Babayigit, A., D'Haen, J., D'Olieslaeger, L., Ethirajan, A., Verbeeck, J., Manca, J., Mosconi, E. and Angelis, F.D., 2015. Intrinsic thermal instability of methylammonium lead trihalide perovskite. Advanced Energy Materials, 5(15), p.1500477.

[10] Correa-Baena, Juan-Pablo, Michael Saliba, Tonio Buonassisi, Michael Grätzel, Antonio Abate, Wolfgang Tress, and Anders Hagfeldt. 2017. Promises and challenges of perovskite solar cells. Science 358, no. 6364, 739-744.

[11] Do Hyung Chun, Sungsoon Kim, Sung Uk Chai, Wook Kim, Wanjung Kim, Jung Hwan Lee, Ryan Rhee, Dukhyun Choi, Jung Kyu Kim, Hyunjung Shin and Jong Hyeok Park. 2019. Grain Boundary Healing of Organic–Inorganic Halide Perovskites for Moisture Stability. Nano letters 19, no. 9, 6498-6505.

[12] Dunfield, S. P., Moore, D. T., Klein, T. R., Fabian, D. M., Christians, J. A., Dixon, A. G., Dou. B., Ardo. S., Beard. M.C., Shaheen S.E., Berry. J. J. and van Hest, M. F.A.M. 2018. Curtailing perovskite processing limitations via lamination at the perovskite/perovskite interface. ACS Energy Letters, 3(5), 1192-1197.

[13] Guichuan Xing, Nripan Mathews, Shuangyong Sun, Swee Sien Lim, Yeng Ming Lam, Michael Grätzel, Subodh Mhaisalkar and Tze Chien Sum. 2013. Long-Range Balanced Electron- and Hole-Transport Lengths in Organic-Inorganic $CH_3NH_3PbI_3$. Science, 342, 344-347.

[14] Han G., Hadi, H. D., Bruno, A., Kulkarni, S. A., Koh, T. M., Wong, L. H., Soci. C., Mathews. N., Zhang. S. and Mhaisalkar, S. G. 2018. Additive selection strategy for high performance perovskite photovoltaics. The Journal of Physical Chemistry C, 122 (25), 13884-13893.

[15] Hao Feng, Constantinos C. Stoumpos, Duyen Hanh Cao, Robert PH Chang, and Mercouri G. Kanatzidis. 2014. Lead-free solid-state organic–inorganic halide perovskite solar cells. Nature photonics 8, 6, 489-494.

[16] Hao Feng, Constantinos C. Stoumpos, Robert PH Chang, and Mercouri G. Kanatzidis. 2014. Anomalous band gap behavior in mixed Sn and Pb perovskites enables broadening of absorption spectrum in solar cells. Journal of the American Chemical Society 136, no. 22 (2014): 8094-8099.

[17] Hui-Seon Kim, Chang-Ryul Lee, Jeong-Hyeok Im, Ki-Beom Lee, Thomas Moehl, Arianna Marchioro, Soo-Jin Moon, Robin Humphry-Baker, Jun-Ho Yum, Jacques E. Moser, Michael Gra¨tzel and Nam-Gyu Park. 2012. Lead Iodide Perovskite Sensitized All-Solid-State Submicron Thin Film Mesoscopic Solar Cell with Efficiency Exceeding 9%. Scientific Reports, 2, 591.

[18] Jaeki Jeong, Minjin Kim, Jongdeuk Seo, Haizhou Lu, Paramvir Ahlawat, Aditya Mishra, Yingguo Yang, Michael A Hope, Felix T Eickemeyer, Maengsuk Kim, Yung Jin Yoon, In Woo Choi, Barbara Primera Dareich, Seung Ju Choi, Yimhyun Jo, Jun Hee Lee, Bright Walker, Shaik M Zakeeruddin, Lyndon Emsley, Ursula Rothlisberger, Anders Hagfeldt, Dong Suk Kim, Michael Gratzel and Jin Young Kim. 2021. Pseudo-halide anion engineering for α-$FAPbI_3$ perovskite solar cells. Nature 592, no. 7854, 381-385.

[19] John S. Mangum, San Theingi, Myles A. Steiner, William E. McMahon, and Emily L. Warren. 2021. Development of High-Efficiency GaAs Solar Cells Grown on Nanopatterned GaAs Substrates. Crystal Growth & Design, 21, 10, 5955-5960.

[20] Jokar Efat, Cheng-Hsun Chien, Cheng-Min Tsai, Amir Fathi, and Eric Wei-Guang Diau. 2019. Robust tin-based perovskite solar cells with hybrid organic cations to attain efficiency approaching 10%. Advanced materials 31, no. 2, 1804835.

[21] Juarez-Perez, Emilio J., Zafer Hawash, Sonia R. Raga, Luis K. Ono, and Yabing Qi. 2016. Thermal degradation of $CH_3NH_3PbI_3$ perovskite into NH_3 and CH_3I gases observed by coupled thermogravimetry–mass spectrometry analysis. Energy & environmental science 9, no. 11, 3406-3410.

[22] Kopacic, I., Friesenbichler, B., Hoefler, S.F., Kunert, B., Plank, H., Rath, T. and Trimmel, G., 2018. Enhanced performance of germanium halide perovskite solar cells through compositional engineering. ACS Applied Energy Materials, 1(2), pp.343-347

[23] Kour Ravinder, Sandeep Arya, Sonali Verma, Jyoti Gupta, Pankaj Bandhoria, Vishal Bharti, Ram Datt, and Vinay Gupta. 2019. Potential substitutes for replacement of lead in perovskite solar cells: A review. Global Challenges 3, no. 11, 1900050.

[24] L. Sowjanya Pali, Jeetendra Kumar Tiwari, Nasir Ali, Subhasis Ghosh, Kanwar S. Nalwa and Ashish Garg. 2021. Development of MoO_3/Au/MoO_3 top transparent conducting electrode for organic solar cells on opaque substrates. Energy Technology, 2100689.

[25] Lee, Jong Woo, Haejun Yu, Kisu Lee, Sohyeon Bae, Jungwon Kim, Gi Rim Han, Doyk Hwang, Seong Keun Kim, and Jyongsik Jang. 2019. Highly crystalline perovskite-based photovoltaics via two-dimensional liquid cage annealing strategy. Journal of the American Chemical Society 141, no. 14, 5808-5814.

[26] Lee, J. W.; Kim, D. H.; Kim, H. S.; Seo, S. W.; Cho, S. and M.; Park, N. G 2015. Formamidinium and Cesium Hybridization for Photo- and Moisture-Stable Perovskite Solar Cell. Adv. Energy Mater. 5, 1501310.

[27] Leguy, A.M., Hu, Y., Campoy-Quiles, M., Alonso, M.I., Weber, O.J., Azarhoosh, P., Van Schilfgaarde, M., Weller, M.T., Bein, T., Nelson, J. and Docampo, P. 2015. Reversible hydration of $CH_3NH_3PbI_3$ in films, single crystals and solar cells. Chemistry of Materials, 27 (9), pp.3397-3407.

[28] Li Wenzhe, Cuiling Zhang, Yunping Ma, Chong Liu, Jiandong Fan, Yaohua Mai, and Ruud EI Schropp. 2018. In situ induced core/shell stabilized hybrid perovskites via gallium (iii) acetylacetonate intermediate towards highly efficient and stable solar cells. Energy & Environmental Science 11, no. 2, 286-293.

[29] Ling Hong, Huifeng Yao, Yong Cui, Ziyi Ge and Jianhui Hou. 2020. Recent advances in high-efficiency organic solar cells fabricated by eco-compatible solvents at relatively large-area scale. APL Materials 8, 120901.

[30] Liu Gengling, Cong Liu, Zhuojia Lin, Jia Yang, Zengqi Huang, Licheng Tan, and Yiwang Chen. 2020. Regulated crystallization of efficient and stable tin-based perovskite solar cells via a self-sealing polymer. ACS applied materials & interfaces 12, no. 12, 14049-14056.

[31] Long, Mingzhu, Tiankai Zhang, Wangying Xu, Xiaoliang Zeng, Fangyan Xie, Qiang Li, Zefeng Chen et al. 2017. Large-grain formamidinium PbI3–xBrx for high-performance perovskite solar cells via intermediate halide exchange. Advanced Energy Materials 7, no. 12, 1601882.

[32] Min Chen, Ming-Gang Ju, Hector F. Garces, Alexander D. Carl, Luis K. Ono, Zafer Hawash, Yi Zhang, Tianyi Shen, Yabing Qi, Ronald L. Grimm, Domenico Pacifici, Xiao Cheng Zeng, Yuanyuan Zhou and Nitin P. Padture. 2019. Highly stable and efficient all-inorganic lead-free perovskite solar cells with native-oxide passivation. Nature communications 10, 1, 1-8.

[33] Mingyu Jeong, In Woo Choi, Eun Min Go, Yongjoon Cho, Minjin Kim, Byongkyu Lee, Seonghun Jeong, Yimhyun Jo, Hye Won Choi, Jiyun Lee, Jin-Hyuk Bae, Sang Kyu Kwak, Dong Suk Kim and Changduk Yang. 2020. Stable perovskite solar cells with efficiency exceeding 24.8% and 0.3-V voltage loss. Science 369, no. 6511, 1615-1620.

[34] Nie, W.; Blancon, J. C.; Neukirch, A. J.; Appavoo, K.; Tsai, H.; Chhowalla, M.; Alam, M. A.; Sfeir, M. Y.; Katan, C.; Even, J, Tretiak, S.; Crochet. J.J.; Gupta. G. and Mohite. A.D. 2016. Light-activated Photocurrent Degradation and Self-healing in Perovskite Solar Cells. Nat. Commun. 7, 11574.

[35] Noel Nakita K., Samuel D. Stranks, Antonio Abate, Christian Wehrenfennig, Simone Guarnera, Amir-Abbas Haghighirad, Aditya Sadhanala et al. 2014. Lead-free organic–inorganic tin halide perovskites for photovoltaic applications. Energy & Environmental Science 7, no. 9. 3061-3068.

[36] Park, N. G.; Gratzel, M.; Miyasaka, T.; Zhu, K.; Emery, K. 2016. Towards Stable and Commercially Available Perovskite Solar Cells. Nat. Energy, 1, 16152.

[37] Pengcheng Wang, Fengzhu Li, Ke-Jian Jiang, Yanyan Zhang, Haochen Fan, Yue Zhang, Yu Miao, Jin-Hua Huang, Caiyan Gao, Xueqin Zhou, Fuyi Wang, Lian-Ming Yang, Chuanlang Zhan and YanLin Song. 2020. Ion exchange/insertion reactions for fabrication of efficient methylammonium tin iodide perovskite solar cells. Advanced Science 7, no. 9, 1903047.

[38] Salado Manuel, Laura Calio, Rüdiger Berger, Samrana Kazim, and Shahzada Ahmad. 2016. Influence of the mixed organic cation ratio in lead iodide-based perovskite on the performance of solar cells. Physical Chemistry Chemical Physics 18, no. 39, 27148-27157.

[39] Saliba Michael, Taisuke Matsui, Ji-Youn Seo, Konrad Domanski, Juan-Pablo Correa-Baena, Mohammad Khaja Nazeeruddin, Shaik M. Zakeeruddin, Wolfgang Tress, Antonio Abate, Anders Hagfeldy and Michael Gratzel. 2016. Cesium-containing triple cation perovskite solar cells: improved stability, reproducibility and high efficiency. Energy & environmental science 9, no. 6, 1989-1997.

[40] Samuel D. Stranks, Giles E. Eperon, Giulia Grancini, Christopher Menelaou, Marcelo Alcocer, Tomas Leijtens, Laura M. Herz, Annamaria Petrozza, and Henry J. Snaith. 2013. Electron-Hole Diffusion Lengths Exceeding 1 Micrometer in an Organometal Trihalide Perovskite Absorber. Science, 342, 341-344.

[41] Stefaan De Wolf, Jakub Holovsky, Soo-Jin Moon, Philipp Löper, Bjoern Niesen, Martin Ledinsky, Franz-Josef Haug, Jun-Ho Yum, and Christophe Ballif. 2014. Organometallic Halide Perovskites: Sharp Optical Absorption Edge and Its Relation to Photovoltaic Performance. The Journal of Physical Chemistry Letters 5, 6, 1035-1039.

[42] Thomson, S. 2018. Observing Phase Transitions in a Halide Perovskite Using Temperature Dependent Photoluminescence Spectroscopy. Edinburgh Instruments.

[43] Tianhao Wu, Zhenzhen Qin, Yanbo Wang, Yongzhen Wu, Wei Chen, Shufang Zhang, Molang Cai, Songyuan Dai, Jing Zhang, Jian Liu, Zhongmin Zhou, Xiao Liu, Hiroshi Segawa, Hairen Tan, Qunwei Tang, Junfeng Fang, Yaowen Li, Liming Ding, Zhijun Ning, Yabing Qi, Yiqiang Zhang and Liyuan Han. 2021. The Main Progress of Perovskite Solar Cells in 2020–2021. Nano-Micro Lett. 13, 152.

[44] Turren-Cruz, Silver-Hamill, Anders Hagfeldt, and Michael Saliba. 2018. Methylammonium-free, high-performance, and stable perovskite solar cells on a planar architecture. Science 362, no. 6413, 449-453.

[45] Wang, Shenghao, Yan Jiang, Emilio J. Juarez-Perez, Luis K. Ono, and Yabing Qi. 2016. Accelerated degradation of methylammonium lead iodide perovskites induced by exposure to iodine vapour. Nature Energy 2, no. 1, 1-8.

[46] Yang Woon Seok, Jun Hong Noh, Nam Joong Jeon, Young Chan Kim, Seungchan Ryu, Jangwon Seo, and Sang Il Seok. 2015. High-performance photovoltaic perovskite layers fabricated through intramolecular exchange. Science 348, no. 6240, 1234-1237.

[47] Yang, Yuqian, Jihuai Wu, Xiaobing Wang, Qiyao Guo, Xuping Liu, Weihai Sun, Yuelin Wei, Yunfang Huang, Zhang Lan, Miaoliang Huang, Jianming Lin, Hongwei Chen and Zhanhua Wei. 2020. Suppressing vacancy defects and grain boundaries via Ostwald ripening for high-performance and stable perovskite solar cells. Advanced Materials 32, no. 7, 1904347.

[48] Yi, Ahra, Sangmin Chae, Hanbin Lee, and Hyo Jung Kim. 2019. The synergistic effect of cooperating solvent vapor annealing for high-efficiency planar inverted perovskite solar cells. Journal of Materials Chemistry A 7, no. 48, 27267-27277.

[49] Yun, Jae S., Anita Ho-Baillie, Shujuan Huang, Sang H. Woo, Yooun Heo, Jan Seidel, Fuzhi Huang, Yi-Bing Cheng, and Martin A. Green. 2015. Benefit of grain boundaries in organic–inorganic halide planar perovskite solar cells. The journal of physical chemistry letters 6, no. 5, 875-880.

[50] Yun, Jae Sung, Jincheol Kim, Trevor Young, Robert J. Patterson, Dohyung Kim, Jan Seidel, Sean Lim, Martin A. Green, Shujuan Huang, and Anita Ho-Baillie. 2018. Humidity-induced degradation via grain boundaries of $HC(NH_2)_2PbI_3$ planar perovskite solar cells. Advanced Functional Materials 28, no. 11, 1705363.

[51] Zhengchun Wang, Chang Shi, Ziyi Wang, Lan Xiao, Tong Wu, Xueli Yu, Liang Ma, Xiangbai Chen, Jiliang Zhang, Hongwei Lei and Pingli Qin. 2022. Solution-processed $Fe_{2-x}Mg_xO_3$ ternary oxides for interface passivation in efficient perovskite solar cells. Chemical Engineering Journal 441, 136118.

[52] Zhou, Huawei, Yantao Shi, Kai Wang, Qingshun Dong, Xiaogong Bai, Yujin Xing, Yi Du, and Tingli Ma. 2015. Low-temperature processed and carbon-based $ZnO/CH_3NH_3PbI_3/C$ planar heterojunction perovskite solar cells. The Journal of Physical Chemistry C 119, no. 9. 4600-4605

Chapter 16

Catalytic Efficacy of Terpyridine Complexes in Carbon Dioxide Reduction

Venkata Srujana B., Rajarshi Sarkar*

Department of Chemistry, School of Advanced Sciences, VIT-AP University

Correspondence: rajarshi.sarkar@vitap.ac.in

ABSTRACT

The development of novel energy storage technologies is very crucial to address the challenges that prevent the broad use of renewable energy sources. Currently, two methodologies—CO_2 reduction, and the conversion of CO_2 into compounds with higher energy, and oxidation/reduction of water, or the creation of H_2 and O_2 respectively—are at the centre of significant research to transfer electrical and/or solar energy into chemical energy. To accomplish these goals, numerous photo-, electro-, and photoelectrochemical techniques have been developed, typically utilizing transition metal complexes as catalysts. With the aid of transition metal complexes, structure-activity relationships have been understood and electrocatalytic systems has been improved. Polypyridine ligands, such as bipyridine (bpy) and terpyridine (tpy), are common ligands in coordination chemistry and molecular catalysis because they frequently produce stable, well-defined complexes. They have been studied so frequently in relation to CO_2 electroreduction. Catalytic efficacy of metal-terpyridine complexes in CO_2 reduction will be discussed in this book chapter.

To overcome the obstacles facing the widespread usage of renewable energies the advancement of novel energy storage technologies is absolutely imperative.[1] Two methodologies at present areat the heart of intense exploration to convert electrical and/or solar energy into chemical energy: CO_2 reduction, or the conversion of CO_2 into compounds with higher energy, andoxidation/reduction of water, or of H_2 and O_2production respectively.[2-6] However, because carbon-based fuels have higher volumetric and gravimetric densities, reducing carbon dioxide (CO_2) to them is more desirable.[7-8]This conversion is carried out either directly by producing methanol, formic acid, and higher hydrocarbons, or *via* indirect method by producing carbon monoxide to utilized as a chemical feedstock for the Fischer-Tropsch process, which produces alkanes.[9] Additionally, reducing CO_2 has the benefit of creating a global carbon-neutral energy system that integrates with existing infrastructure and streamlines energy transportation.[10] A wide variety of photo-, electro-, and photoelectrochemical strategies have been devised to achieve these objectives, frequently using transition metal complexes as catalysts.[11]Structure-activity interactions can be better understood and electrocatalytic systems can be optimized with the help of transition metal complexes.[12] As they typically produce stable, well-defined complexes, several polypyridine ligands, including bipyridine (bpy) and terpyridine (tpy), are common ligands in coordination chemistry and molecular catalysis.[13] They have so frequently been researched in the context of CO_2 electroreduction.

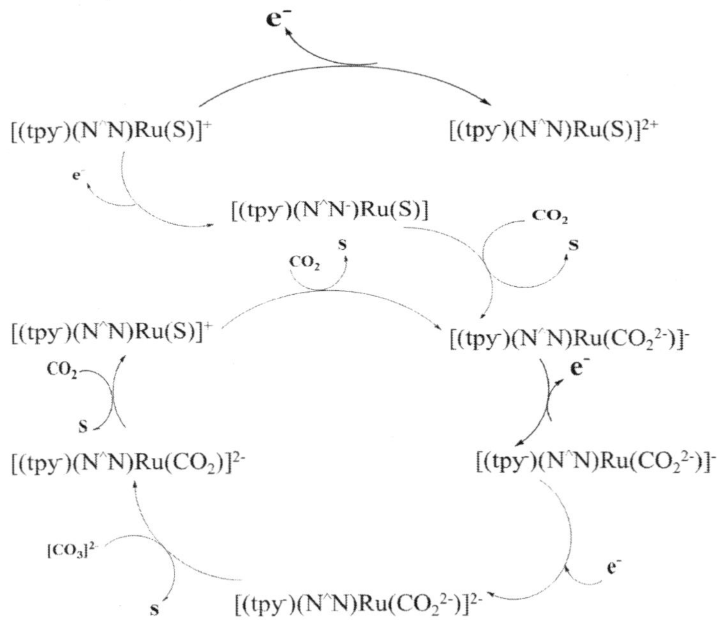

Figure 1. Schematic representation of the electrocatalytic CO_2 reduction using a $[Ru(tpy)(N^\wedge N)(S)]^{2+}$ complex as catalyst (N^N denotes a bidentate ligand, S represents a coordinated solvent molecule).

In their seminal work Tanaka *et al*. electrochemically reduced CO_2 by employing a heteroleptic complex $[Ru(tpy)(bpy)(CO]^{2+}$.[14] But the catalyst functioned at a relatively negative potential and proceeded with a sluggish rate which is usual for the majority of the complexes reported in literature. The aqua complexes, **1**(Figure 2; L = OH_2) which were found to be multifaceted single-site catalysts for the oxidation of water, were explored with respect to their ability to reduce CO_2 to CO electrocatalytically.[15] The fundamental catalytic cycle of reduction of CO_2 to CO as demonstrated by Meyer et al., is presented in Figure 1.[16] Initially, a precursor is converted into a solvato complex, and aterwards this solvato complex enters the catalytic cycle through two one-electron reduction process involving the bidentate

NN ligand and tpy moiety. The catalytic performance of the metal-complex was discovered to hugely depend upon the nature of the ligand. The catalytic rate constants (k_{cat}) of the complex **1b** with an *N*-heterocyclic carbene (NHC) ligand in CO_2-saturated solutions, were found to be 5.5 and 19.1 s^{-1}, respectively. The enhancedcatalytic activity of the doubly-reduced complex **1b**was attributed to the improved labilty of the solvato ligand. It is also noteworthy, hydride complex $[Ru(tpy)(bpy)H]^+$ was electrogenerated from **1a** *in situ*, catalyzes reduction of acetone to 2–propanol.[17] Variety ofmetal complexes with structural modification at

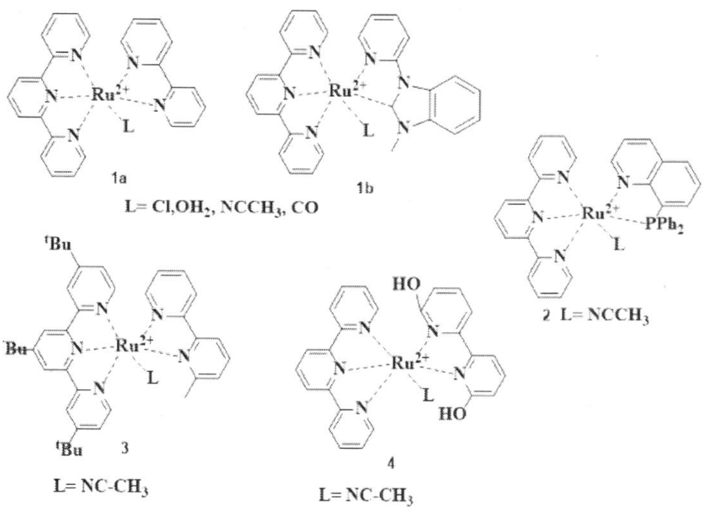

Figure 2. Schematic representation of complexes used as catalysts in the context of electrocatalytic CO_2 reduction.

the tpy and/or the bpy moieties has been employed to electrocatalytically reduce CO_2 in accordance with Meyer's foundational investigations. The electron density at the Ru(II) core can be directly influenced by the electron-donating or –withdrawing nature of the substituents, making the bidentate ligand in particular vulnerable to electronic modification. This consequently regulates the lability of the monodenate ligand L, which is related to its ability to bind CO_2 and eventually to its electrocatalytic activity. Ott *et al.* postulated a variety of mechanistic pathways which could be followed under different conditions can result in a CO_2reduction.[18] For instance, complex **2** with an N-P ligand was used by Masaoka *et al.* to reduce CO_2 at substantially lower overpotentials than complex **1a**.[19] Mechanistic investigations suggested that the departure of the solvato ligand and the fixation of CO_2 were facilitated by the strong P-donor moiety in the trans-position to it. In connection with this, a CH_3-substituent on the bpy unit in **3** significantly enhanced the electrocatalytic activity (Figure 2).[20-21] An initiatoryelectron transfer followed by a chemical reaction, and a subsequent electron transfer (ECE-type mechanism) are the characteristics this catalytic cycle. The dissociation of the solvato ligand emerged more advantageous due to steric hindrance. In this instance, the CO_2 binding has already taken place in the singly reduced form, allowing entry into a catalytic cycle that the parent complex **1a** is unable to access.

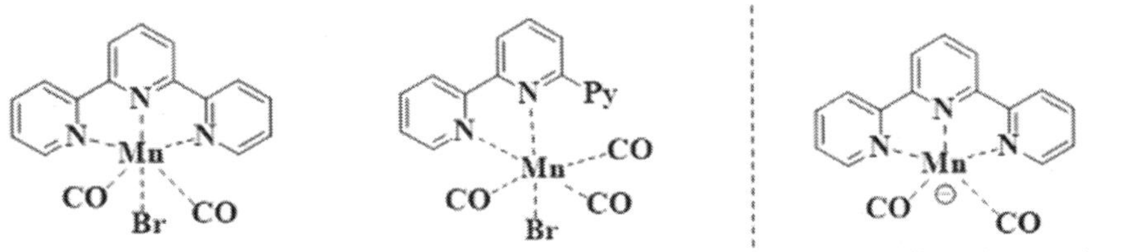

5 6 Catalytically active species

Figure 3. Schematic representation of the Mn(I)–tpy complexes 5&6 which both yield the catalytically active $[Mn(tpy)(CO)_2]^-$ species owing to two consecutive one-electron reductions.

As a result, a remarkably low overpotential for the CO_2 reduction of 0.47 V was discovered, along with a nearly quantitative coulombic efficiency (for correlation, the overpotential of **1a** was 0.87 V under the same operating conditions).[22] It should be noted that stronger electron-donating substituents do not always result in increased electrocatalytic activity. Binding of CO_2 to the metal centre was found to result in weak catalytic activity and quick deactivation because ensuing one electron reduction produced an inert Ru(II)-carbonyl molecule.

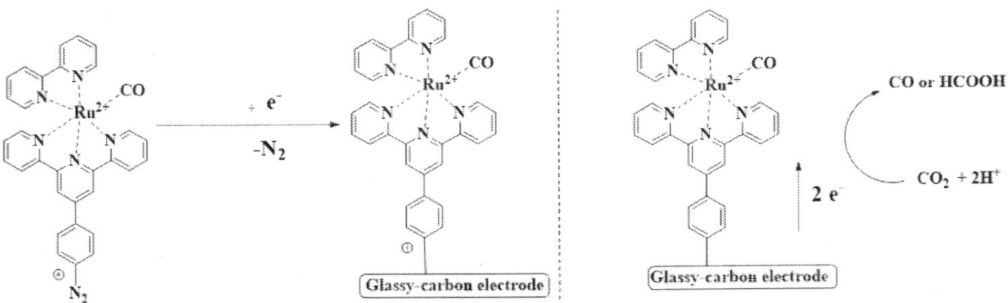

Figure 4. Schematic representation of the electrochemical functionalization of glassy-carbon electrodes and their utilization in the electrocatalytic CO_2 reduction.

Fontecave *et al.* developed different homoleptic $[M(tpy)2]^{2+}$ complexes (M = Fe, Mn, Ni, Co, Cu, and Zn) in order to substitute inexpensive and abundant first-row transition metal ions for the rare and expensive heavy-metal centres in those complexes.[23] However, the complexes comprising of Co(II) and Ni(II) were found to be effective in the electrocatalytic reduction of CO_2. The complex $[Ni(tpy)2]^{2+}$demonstrated a good selectivity for reducing CO_2 to CO while inhibiting the hydrogen evolution reaction. The high valent metal complex prevents the synthesis of Ni-hydride species, which facilitates for the formation of H_2. On the other hand, the lowvalent $[Co(tpy)2]^{2+}$produced a mixture of CO and H_2. By varying the applied overpotential, it is possible to regulate the ratio of CO/H_2: While CO formation was favourable at low potentials, the $CO{:}H_2$ ratio fell considerably as the potential increased (a ratio of 1 was reachedat –2.1 V). The CO/H_2 ratio can be changed by chemically altering the tpy ligand as well, as reported by the same authors.[24]

Machan and coworkers synthesized Mn(I)-tpy mono-complexes suitable for catalyzing electrocatalytic reduction of CO_2 by employing a different approach.

Similar compounds based on the heavy homolog of manganese, Re, were already investigated in 1990s.[25] It was observed that in homogeneous conditions the complex [Re(ph-tpy)$(CO)_3$Cl] assists the photo- and electroreduction of CO_2. Upon one-electron reduction, the Mn(I)-tpy complexes **5&6** produce a dimeric Mn(0) species which subsequently leads to the catalytically active $[Mn(tpy)CO)_2]^-$ complex (Figure 3).[26] Low catalytic performance of this catalyst in comparison with the more active Mn(I)-bpy speciesin generation of CO from CO_2 was attributed to an unsuitable ligand architecture[27] Along with the previously mentioned homogeneous catalysts, heterogeneous versions have also been employed in light of the synthesis of

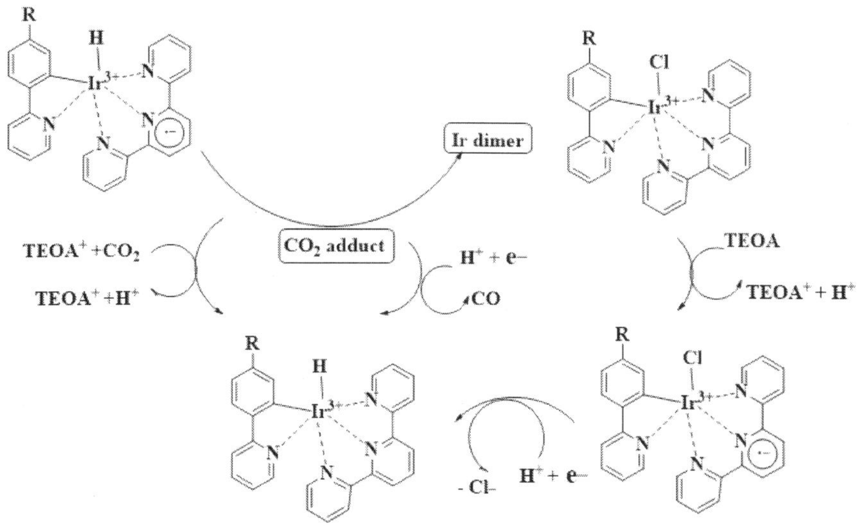

Figure 5. Schematic representation of the proposed mechanism for the photocatalytic CO_2 reduction using the heteroleptic Ir(III) complexes as photocatalyst

higher-energy organic compounds from CO_2*via* electrochemical reduction. This method involves immobilizing homogeneous catalysts onto suitable solid supports and when the support is conductive, the immobilization results in a chemically altered electrode that can be used for electrochemical catalysis. Nanocrystalline TiO_2 and glassy-carbon electrodes have been used in this situation specifically. Abruña*et al.* in their pioneering work coated the previous complex with thin films of metallopolymer. Electropolymerization of a Co(II)-tpy *bis*complex functionalized with vinyl generated these films.[28] In theinitial investigation, formic acid was produced in CO_2-saturated dimethyl formamide at -0.90 V. Furthermore, it was reported high that surface coverages allowed CO_2 to permeate and, as a result, overall catalytic activity was reduced. Additionally, the same authors investigated the reduction of CO_2 in aqueous media utilizing carbon electrodes wrapped in thin films made from different $[M(tpy)2]^{2+}$ complexes (M = Ni, Cr, Fe, Co, Ru *etc.*) with vinyl-functionalization.[29] The electrodes coated with first-row transition metal-complexes showed the highest catalytic activity. The electropolymerized films of $[Cr(tpy)2]^{2+}$ displayed a current efficiency of almost 87% and a TON values of greater than 15,000 were attained.

Tanaka *et al.* reported the electrochemical reduction of diazoniumfunctionalized Ru(II) complexes to produce multi-layer films on glassy-carbon (GC) electrodes (Figure 4).[30]The prepared electrodes were employed in the reduction of CO_2 to produce CO or formic acid elctrocatalytically. In a comparable study, vacant tpy sites were functionalized on a GC electrode with Co(II) ions by Fontecave *et al.*[31]The CO generation from CO_2 was only moderate in comparison to the previous complex. Single-walled carbon nanotubes (SWCNTs) were functionalized by McQueen and Goldsmith with $[Co(tpy)2]^{2+}$*via* remaining pyrene units that were attached with the SWCNTs supramolecularly, i.e., creating strong stacking interactions.[32]It has been demonstrated that nanostructured dye-sensitized TiO_2 photoelectrodes can be used to populate long-lived charge-separated states (CSSs), which are necessary for multi-electron CO_2 reduction.

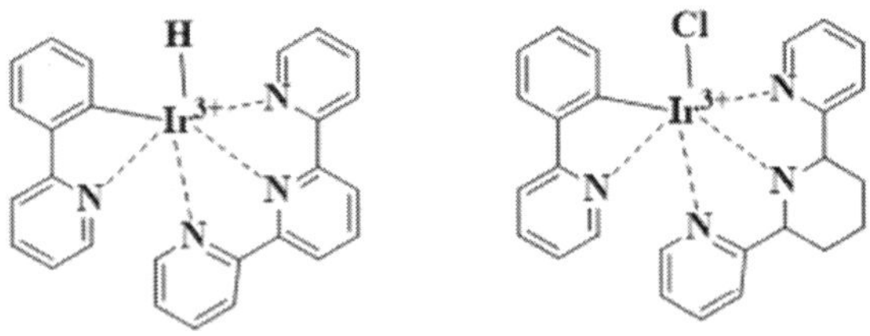

$Ir(tpy)(bpy)Cl^{2+}$, **7** $[Ir(red-tpy)(bpy)H]^{2+}$, **8**

Figure 6. Schematic representation of the dicationic Ir(III) photocatalyst 7 and the ligand-reduced hydride complex 8, which formed in course of the photocatalytic CO_2 reduction.

In light of this, Meyer *et al.* used a phosphonate moiety to bind Ott's Ru(II) catalysed (Figure 1) onto nanocrystalline TiO_2. Additionally, above the complex layer a protective overlayer of NiO was deposited. This heterogeneous construction was used in the electrocatalytic CO_2 reduction.[31] At a potential of 1.65 V vs. Fc+/Fc, CO was formed; a maximum TON value of 237 was attained over the course of 4 hours. The molecular structure of the catalyst was unaffected by the prolonged electrocatalytic run, confirmed by the XPS studies.Additionally, the heterogeneous catalyst was significantly more effective than its homogeneous counterpart: In these trials, an extremely low TON value of 3 and a Faradaic efficiency of 62.5% were noted. The reason for the low performance was presumed to be the diffusion constraint of the homogeneous catalyst as well as an insufficient interaction of the electrode with the catalyst within a predecided duration. Therefore, to achieve strong electrocatalytic performances the immobilization of catalysts onto electrode surface is found to be quite advantageous.[32] Chauvin et al. immobilized catalyst Mn(I) **6**using its lateral phosphonate moieties (Figure 3) and photosensitizer $[Ru(bpy)_3]^{2+}$ on TiO_2.[33]The photoelectrochemical conversion of CO_2 to formic acid using the coassembled system was found to be effective and selective (a quantum yield of 0.17% was reported). According to electrochemical studies, the conduction band of the semiconductor served as the pathway for the transport of electrons from the photo-excitated Ru(II) core to the Mn(I) site. The monomeric one-electron reduced Mn(0) species was identified to be the catalytically active. It is worth noting that dimeric Mn(0)-Mn(0) entities were responsible in the production of CO under homogenous circumstances as a precatalyst.[34] Reisner and coworkers synthesized a $[Co(tpy)2]^{2+}$ complexes functionalized with phosphonate onto a light-harvesting *p*-type Si electrode, mixed with a layer of mesoporous TiO_2, in order to completely eliminate rare and expensive metals.[35]A different mechanistic pathway for heterogeneous catalysis from the one that holds true in homogeneous settings was discovered using spectroelectrochemical investigations. The heterogeneous setup allowed for the light-driven catalysis to continue not just in aqueous organic solvents but also in aqueous media, at a lower potential and with higher activity.[36]

In addition to the previously described (light-driven) electrocatalytic reduction of CO_2, the photocatalytic reduction has also garnered a lot of attention.The heteroleptic Ir(III)-tpy complexes with the common formula $[Ir(tpy)(ppy)L]^+$(L = Cl or CN) is worth mentioning among the numerous homogeneous[37-38] and heterogeneous[39]photocatalytic systems reported in the literature. Ishitani *et al.* looked at the complexes **7** (Figure 6) in terms of how well they could use triethanolamine as a sacrifice donor to photochemically decrease CO_2.[40]It was discovered that **8** was the most effective candidate among the complexes under consideration for the photoreduction of CO_2 to CO (TON = 50, = 0.21).The photocatalytic cycle produced the two-electron reduced identical penta-coordinated $[Ir(tpy)(ppy)]^0$*C-trans* and the *C-cis* hydride isomer, to which CO_2 bound in a *C-trans*-fashion. But only the *C-trans*-$[Ir(tpy)(ppy)H]^+$ interacted with CO_2 in the ground state to produce an adduct.[41] Special focus was given by Bernhard *et al.* on the photophysical characteristics of **8** (and a similar complex with CN– as monodentate ligand).[42]These complexes showed outstanding photo-stability, especially in strongly coordinating solvents, which is a requirement for utilizing them as photocatalysts for the CO_2 reduction.[43]The dicationic complex $[Ir(tpy)(bpy)Cl]^{2+}$ (**7**, Figure 6) was also used by Sato *et al.* as a

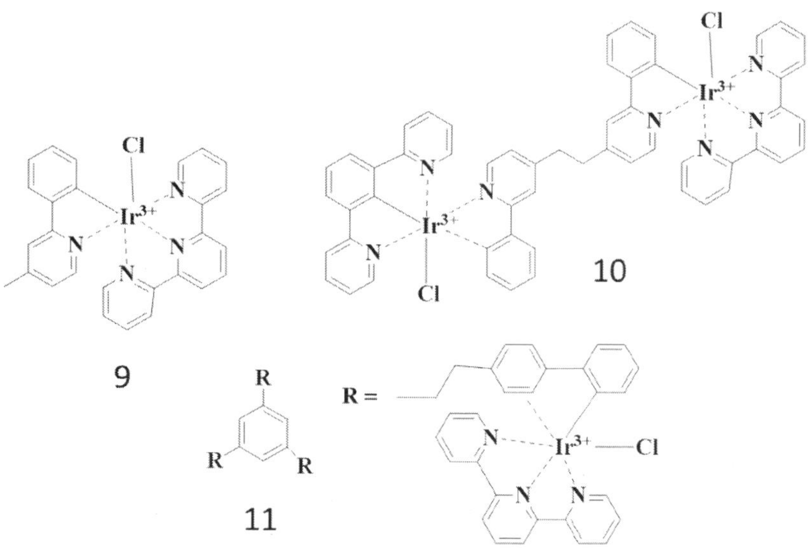

Figure 7. Schematic representation of the mono- (9), di- (10) and trinuclear Ir(III) complexes employed as photocatalyst, for the reduction of CO2.

photocatalyst for the reduction of CO_2.[44-45] *In situ* spectroscopic observations confirmed the generation of the hydride species $[Ir(tpy)(bpy)H]^+$, involved in the selective CO synthesis, in the case of its cyclometalated analogues **7**. It was possible to isolate and analyse the hydride complex **8** (Figure 6) when this complex was further ligand-reduced at a longer irradiation time.[46] As a result, this species no longer supported the generation of CO but instead promoted the selective production of formic acid. When preformed **8**was present during the photocatalytic CO_2 reduction (at 436 nm), a quantum yield of 2.1% was attained for the production.[47-48]Rieger *et al.* compared the photocatalytic % performance of mononuclear $[Ir(tpy)(bpy)Cl]^+$ to that of comparable di- and tri-nuclear constructs.[49-50] The set of complexes **9-11** was prepared with this in mind (Figure 7). In the case of **9**, the photocatalytic reduction of CO_2 resulted in CO with the highest quantum efficiency of 18%. The efficiency of the dinuclear complexes **10b**was comparable to that of **10a**. A process of intramolecular quenching, made possible by the small spacer unit, was responsible for the poor efficiency of **10a**. Nonetheless, as evidenced by their TON values of >80, all dinuclear complexes demonstrated increased photo-stability (a TON of 33 was reported for **9**).The estimated TOF values for the photocatalytic activity showed that **9** had the greatest value (TOF = 35 h^{-1}), followed by 37 and 36a (TOF of 21 h^{-1} and 12 h^{-1}, respectively).

A sustainable, closed carbon cycle can still be achieved through electrocatalytic conversion of CO_2 to CO. Various catalysts have been synthesized for this purpose, drawing design inspiration from natural prototypes like CO and formate dehydrogenase enzymes. However, most of the catalysts are prepared from rare and expensive transition metals, lack stability and suffers from low efficiency. Hence this area of research demands attention of the scientific community.

REFERENCE

[1] N. S. Lewis and D. G. Nocera, *Proc. Natl. Acad. Sci.* U. S. A., **2006**, *103*, 15729.

]2] B. Kumar, M. Llorente, J. Froehlich, T. Dang, A. Sathrum, C. P. Kubiak, *Ann. Rev. Phys. Chem.***2012**, *63*, 541.

[3] R. J. Lim, M. Xie, M. A. Sk, J.-M. Lee, A. Fisher, X. Wang, K. H. Lim, *Catal. Today***2014**, *233*, 169.

[3] I. Roger, M. A. Shipman, M. D. Symes, *Nat. Rev. Chem.* **2017**, *1*, 0003.

[4] S. Anantharaj, S. R. Ede, K. Sakthikumar, K. Karthick, S. Mishra, S. Kundu, *ACS Catal.* **2016**, *6*, 8069.

[5] J. Qiao, Y. Liu, F. Hong, J. Zhang, *Chem. Soc. Rev.* **2014**, *43*, 631.

[6] R. Francke, B. Schille, M. Roemelt,*Chem. Rev.* **2018**, *118*, 4631.

[7] T. R. Cook, D. K. Dogutan, S. Y. Reece, Y. Surendranath,T. S. Teets and D. G. Nocera, *Chem. Rev.*, **2010**, *110*, 6474.

[8] H. Arakawa, M. Aresta, J. N. Armor, M. A. Barteau,E. J. Beckman, A. T. Bell, J. E. Bercaw, C. Creutz, E. Dinjus,

D. A. Dixon, K. Domen, D. L. DuBois, J. Eckert, E. Fujita, D. H. Gibson, W. A. Goddard, D. W. Goodman, J. Keller,

G. J. Kubas, H. H. Kung, J. E. Lyons, L. E. Manzer, T. J. Marks, K. Morokuma, K. M. Nicholas, R. Periana, L. Que,

J. Rostrup-Nielson, W. M. H. Sachtler, L. D. Schmidt, A. Sen, G. A. Somorjai, P. C. Stair, B. R. Stults and W. Tumas, *Chem. Rev.*, **2001**, *101*, 953.

[9] M. Rakowski DuBois and D. L. DuBois, *Acc. Chem. Res.*, **2009**, *42*, 1974.

[10] E. E. Benson, C. P. Kubiak, A. J. Sathrum and J. M. Smieja, *Chem. Soc. Rev.*, **2009**, *38*, 89.

[11] A. J. Morris, G. J. Meyer and E. Fujita, *Acc. Chem. Res.*, **2009**,*42*, 1983

[12] C. Kaes, A. Katz and M. W. Hosseini, *Chem. Rev.*, **2000**, *100*, 3553.

[13] J.-M. Save´ant, *Chem. Rev.*, **2008**, *108*, 2348.

[14] H. Nagao, T. Mizukawa, K. Tanaka, *Inorg. Chem.* **1994**, *33*, 3415.

[15] J. J. Concepcion, J. W. Jurss, M. R. Norris, Z. Chen, J. L. Templeton, T. J. Meyer, *Inorg. Chem.* **2010**, 49, 1277.

[16] Z. Chen, C. Chen, D. R. Weinberg, P. Kang, J. J. Concepcion, D. P. Harrison, M. S. Brookhart, T. J. Meyer, *Chem. Commun.* **2011**, 47, 12607.

[17] Z. Chen, C. R. K. Glasson, P. L. Holland, T. J. Meyer, *Phys. Chem. Chem. Phys.* **2013**, *15*, 9503.

[18] T. A. White, S. Maji, S. Ott, *Dalton Trans.* **2014**, *43*, 15028.

[19] S. K. Lee, M. Kondo, G. Nakamura, M. Okamura, S. Masaoka, *Chem. Commun.* **2018**, 54, 6915.

[20] B. A. Johnson, H. Agarwala, T. A. White, E. Mijangos, S. Maji, S. Ott, *Chem. Eur. J.***2016**, 22, 14870.

[21] B. A. Johnson, S. Maji, H. Agarwala, T. A. White, E. Mijangos, S. Ott, *Angew. Chem. Int. Ed.* **2016**, *55*, 1825.

[22] L. Duan, G. F. Manbeck, M. Kowalczyk, D. J. Szalda, J. T. Muckerman, Y. Himeda, E. Fujita, *Inorg. Chem.* **2016**, *55*, 4582.

[23] N. Elgrishi, M. B. Chambers, V. Artero, M. Fontecave, *Phys. Chem. Chem. Phys.* **2014**, *16*, 13635.

[24] N. Elgrishi, M. B. Chambers, M. Fontecave, *Chem. Sci.* **2015**, *6*, 2522.

[26] G. Calzaferri, K. Hädener, J. Li, J. *Photochem. Photobiol. A: Chem.* **1992**, *64*, 259.

[28] C. W. Machan, C. P. Kubiak, *Dalton Trans.***2016**, *45*, 17179.

[29] A. Sinopoli, N. T. La Porte, J. F. Martinez, M. R. Wasielewski, M. Sohail, *Coord. Chem. Rev.* **2018**, *365*, 60.

[30] H. C. Hurrell, A. E. Mogstad, D. A. Usifer, K. T. Potts, H. D. Abruña, *Inorg. Chem.* **1989**, 28, 1080.

[31] J. A. Ramos Sende, C. R. Arana, L. Hernandez, K. T. Potts, M. Keshavarz-K., H. D. Abruña, *Inorg. Chem.* **1995**, *34*, 3339.

[32] Y. Tsukahara, T. Wada, K. Tanaka, *Chem. Lett.* **2010**, *39*, 1134.

[33] N. Elgrishi, S. Griveau, M. B. Chambers, F. Bedioui, M. Fontecave, *Chem. Commun.* **2015**, 51, 2995.

[34] E. W. McQueen, J. I. Goldsmith, J. Am. Chem. Soc. 2009, 131, 17554- 17556.

[35] E. Sundin, M. Abrahamsson, Chem. Commun. 2018, 54, 5289-5298.

[36] T.-T. Li, B. Shan, W. Xu, T. J. Meyer, ChemSusChem 2019, 12, 2402-2408.

[37] L. Le-Quang, M. Stanbury, S. Chardon-Noblat, J.-M. Mouesca, V. Maurel, J. Chauvin, Chem. Commun. 2019, 55, 13598-13601.

[38] J. J. Leung, J. Warnan, K. H. Ly, N. Heidary, D. H. Nam, M. F. Kuehnel, E. Reisner, Nat. Catal. 2019, 2, 354-365.

[39] J. J. Leung, J. A. Vigil, J. Warnan, E. Edwardes Moore, E. Reisner, *Angew. Chem. Int. Ed.* **2019**, *58*, 7697.

[40] W. W. Kramer, C. C. L. McCrory, *Chem. Sci.* **2016**, *7*, 2506.

[41] H. Takeda, O. Ishitani, *Coord. Chem. Rev.* **2010**, *254*, 346.

[42] C. D. Windle, R. N. Perutz, *Coord. Chem. Rev.* **2012**, *256*, 2562.

[43] J. L. White, M. F. Baruch, J. E. Pander, Y. Hu, I. C. Fortmeyer, J. E. Park, T. Zhang, K. Liao, J. Gu, Y. Yan, T. W. Shaw, E. Abelev, A. B. Bocarsly, *Chem. Rev.* **2015**, *115*, 12888.

[44] S. Sato, T. Morikawa, T. Kajino, O. Ishitani, *Angew. Chem. Int. Ed.* **2013**, *52*, 988.

[45] K. Garg, Y. Matsubara, M. Z. Ertem, A. Lewandowska-Andralojc, S. Sato, D. J. Szalda, J. T. Muckerman, E. Fujita, *Angew. Chem. Int. Ed.* **2015**, *54*, 14128.

[46] D. N. Chirdon, W. J. Transue, H. N. Kagalwala, A. Kaur, A. B. Maurer, T. Pintauer, S. Bernhard, *Inorg. Chem.* **2014**, *53*, 1487.

[47] S. Sato, T. Morikawa, *Chem Photo Chem***2018**, *2*, 207.

[48] R. O. Reithmeier, S. Meister, B. Rieger, A. Siebel, M. Tschurl, U. Heiz, E. Herdtweck, *Dalton Trans.* **2014**, *43*, 13259.

Chapter 17

Rhodium Catalyzed Organic Transformations: S Recent Perspective

Sindhu Malisetti, Aparna Pulibandla, Divya Challa, Karri Eswar Brahma Reddy, Shaik Manjirunnisa, Dr. Dripta De Joarder*

Department of Chemistry, School of Advanced Sciences, VIT-AP University, Andhra Pradesh, India

Email: dripta.joarder@vitap.ac.in

ABSTRACT

Transition metal catalyzed reactions have changed the field of organic chemistry significantly. Among the transition metals Rhodium has been particularly useful in academia as well as in industries. Despite being expensive, scientists are turning towards Rh for designing novel catalytic reaction due to its ability to shuttle between three oxidation states *viz*. Rh(I), Rh(II) and Rh(III) giving rise to an array of diverse organic transformations. In this book chapter we will focus on some important Rh catalyzed reactions that have been reported recently in the literature.

INTRODUCTION

Rhodium, despite being expensive has created lot of interest among the scientist community recently as it might be possible to create more environmentally friendly procedures with Rh.[1] This chapter describes important Rhodium catalyzed reactions that have been published recently in the literature.

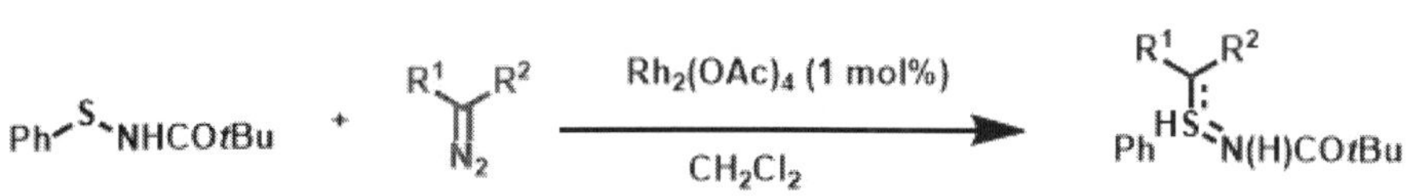

Scheme 1. Catalytic enantioselective S-alkylation of sulfinamides

Dr. Ellman and his team reported a Rhodium catalyzednovel, enantioselective S-alkylation of synthesis of sulfenamides which resulted in sulfilimines.[2] They screened multiple chiral dirhodium catalysts, which are available commercially and established that chiralcatalyst, Rh_2(R-p-Ph-TPCP)$_4$ accomplished best results giving highest enantioselectivity and good yield.

Falck *et. al.*has identified a potent strategy for the functionalization of distant C(sp3)-hydrogens that exploits the well-known regio selectivity of iminyl reactive intermediates.[3] The representative reactions have been summarized in scheme 2. A diverse array of nucleophiles has been utilized. The most acceptable substrates include allylic,propargylic, benzylic,tertiary, and -alkyloxy C(sp3)-hydrogens because of their higher reactivity. The most acceptable substrates include allylic, propargylic, benzylic, tertiary, and -alkyloxy C(sp3)-hydrogens because of their higher reactivity. Unactivated methylenes give low yields and methyls areunreactive.

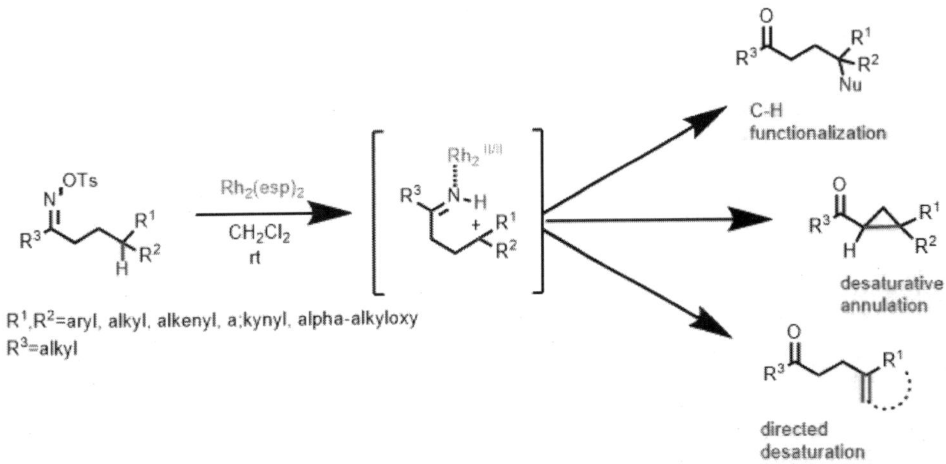

Scheme 2. Remote C(sp3)-H transformations.

Aziridines are useful scaffolds that are abundantin bioactive natural products.[4] Functionalized amines can also be synthesized from aziridines.[5] Trisubstituted aziridines in particular are of utmost significance since they provide access to compounds with a tetrasubstituted carbon centre.[6]For the creation of enantiopure nitrogen-containing compounds, it is therefore of uttermost interest to develop asymmetric catalytic processes for the synthesis of chiral aziridines.

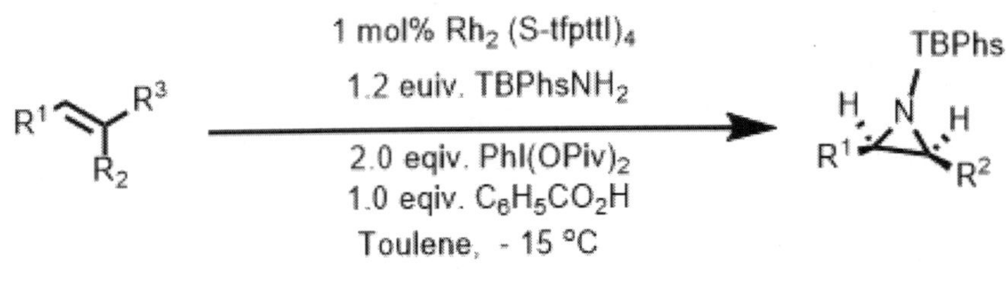

Scheme 3. Enantioselective Intermolecular Aziridinationof Alkenes

Dr. Dauban and his co-workers have demonstrated[7] that highly efficientcatalytic asymmetric alkene aziridination reactions can beperformed with chiral C4-symmetrical dirhodium(II) tetracarboxylatesin the presence of aromatic sulfamates and aniodine(III) oxidant (scheme 3). The process has excellent chemo- and enantioselectivity, with enantiomeric excesses of up to 99%, scalability to the multigram level, wide scope, including mono-, di-, and trisubstituted alkenes, application to late-stage functionalization of complex products, streamlined access to sterically hindered enantiopure benzylic amines.and low catalyst loading.

Anilines are structural scaffolds that are prevalent in functional materials, medicines, agricultural products, and natural products.[8] Hence, it is crucial for synthetic chemists to develop effective techniques for creating molecules with these favoured moieties. Dr. Shi and his colleagues illustrated a rhodium-catalyzed direct amination of phenols with amines without using an activating reagent or a reductant/oxidant.[9]This method is particularly interesting as it provides direct access to substituted anilines with water being the only by product.

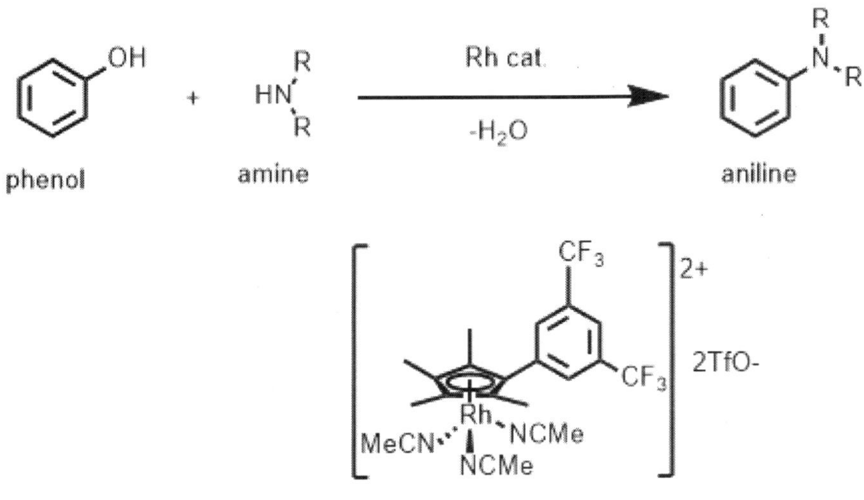

Scheme 4. Catalytic amination of phenols to furnish substituted anilines

1,3-diamines are one remarkable class of structural units that are abundant in drug molecules.[10]Using commercially available NFSI and widely available allylic chemicals catalysed by Rh2 under benign circumstances, Prof. Wang and his team members created a one-step procedure for the effective synthesis of 1,3-diamine compounds.[11] It was shown that internal alkenes, terminal alkenes, and even their mixtures can be used under optimized reaction conditions to obtain the same allylic 1,3-diamine compounds through the modulation of the phenyl substituent of the alkenes.

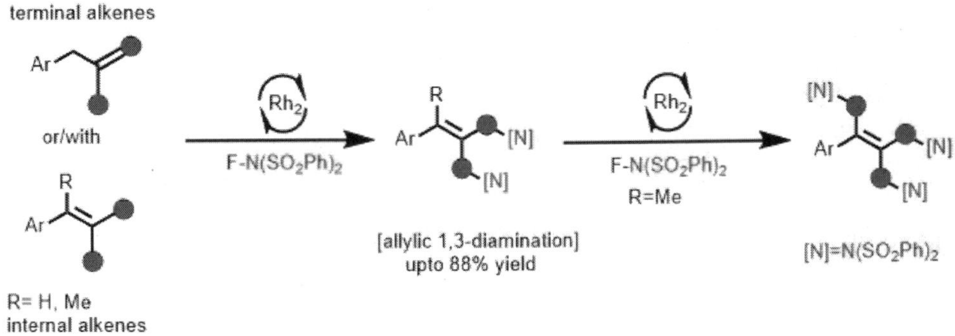

Scheme 5. Rhodium(II)-catalysed direct allylic 1,3-diamination

Many biologically active compounds have key structural motifs that include amino acids with a γ-stereocenter or β,γ -vicinal stereocenters. Dr. Li and his associates published an article[12] describing Rh catalysed regio-, diastereo-, and enantioselective hydroboration ofunactivated alkenes in allyl amides. The methodology is applicable to a broad range of substrates containing a variety of functional groups.

Ma and his colleagues have created a unique mild-condition rhodium-catalyzed asymmetric C-H allylation of indoles with trisubstituted allenes.[13]The method demonstrates exceptional atom economy, Z-selectivity, regioselectivity, high efficiency of chirality transferand outstanding functional group tolerance.

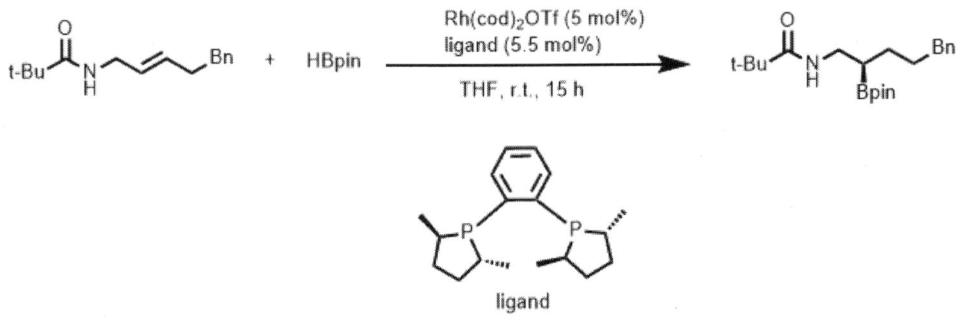

Scheme 6. Rh(I)-catalysed hydroboration of disubstituted alkenes

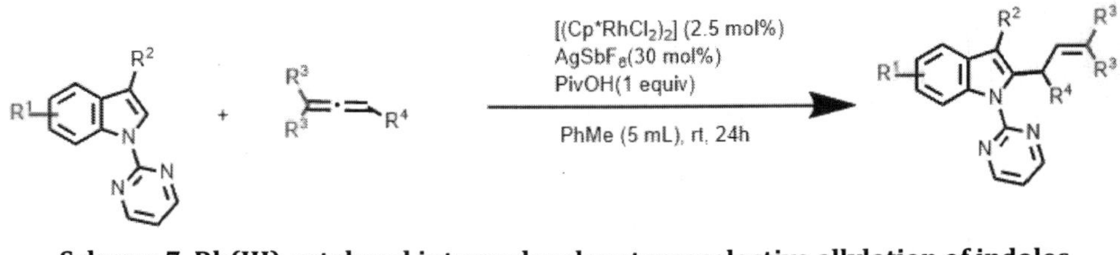

Scheme 7. Rh(III)-catalysed intermolecular stereoselective allylation of indoles

Dr. Zhao and his team members have developed a highly efficient catalyticdeoxygenative reduction of 1,3-diketones using commerciallyavailable rhodium catalyst ([Rh(cod)Cl]$_2$), ligand (Xantphos),and deoxygenating reagent (B$_2$pin$_2$).[14]

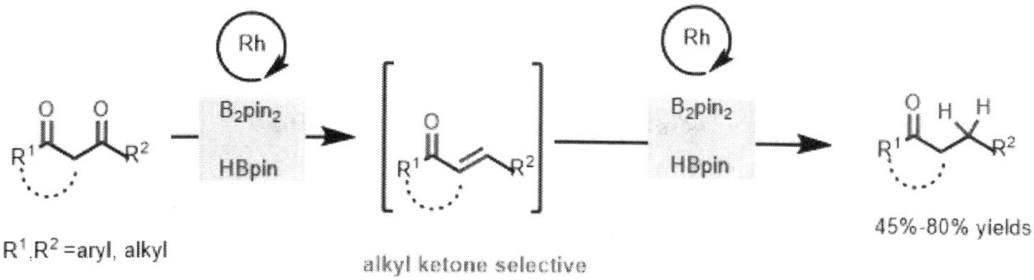

Scheme 8. Rh(I)-catalyzed selective deoxygenativereduction of 1,3-diketones

This methodology exhibits excellent regioselectivity for aliphatic carbonyl reductions over aromatic carbonyl reductions and high chemoselectivity for ketone reductions over olefin and ester reductions. Furthermore, due to the mild reaction conditions, this protocol showed good functional group tolerance and broad substrate range, as well as its application to the synthesis of biologically active molecules.

Dr. Bach and his colleagues have achieved highly chemo- and diastereoselective arene hydrogenation of complex (hetero)aromatic compounds utilising a rhodium(I)-based hydrogenation technique.[15] As delineated in scheme 9, the ability to create up to eight stereogenic centres in one step with a known outcome is a major feature of this technique. A number of stereospecific derivatization tests were carried out, and there is a significant functional group tolerance.

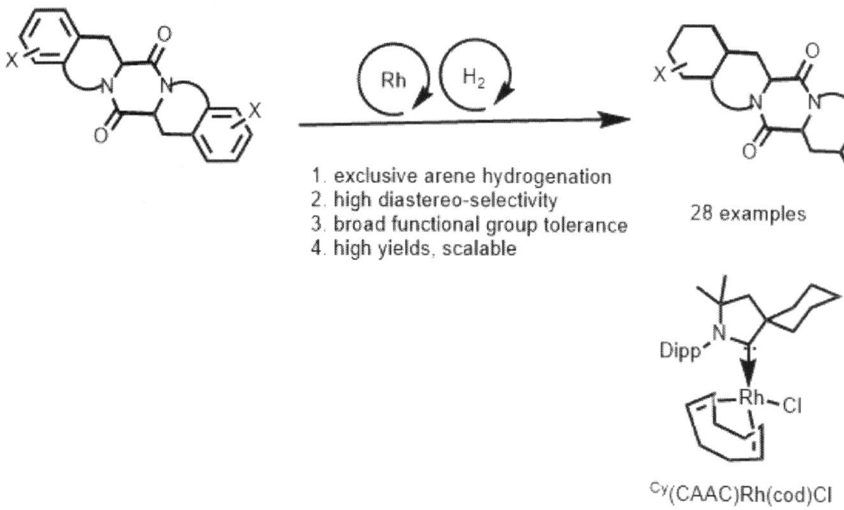

Scheme 9. Diastereoselective Rhodium-catalyzed Arene Hydrogenation

Using Rh(I) catalyst, Dou has successfully established the catalytic asymmetric conjugate pyridylation process, which is a modification of Hayashi-Miyaura reaction.[16]

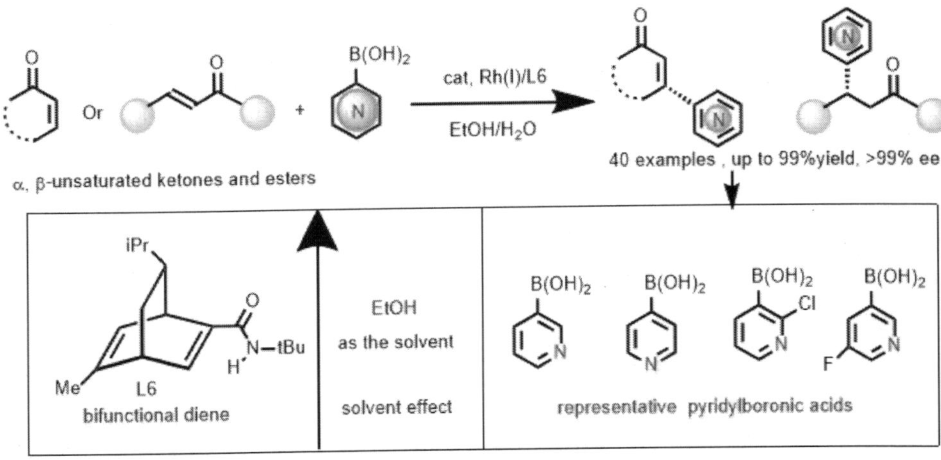

Scheme 10. Hayashi-Miyaura reaction with pyridyl boronic acid

It was discovered that the reaction's performance depended heavily on the bifunctional diene ligand and ethanol solvent, which cooperated to make the catalytic system highly enantioselective and reactive. A wide variety of pyridylboronic acids and,α,β-unsaturated ketones/esters underwent the reaction smoothly proving wide applicability of the reaction.

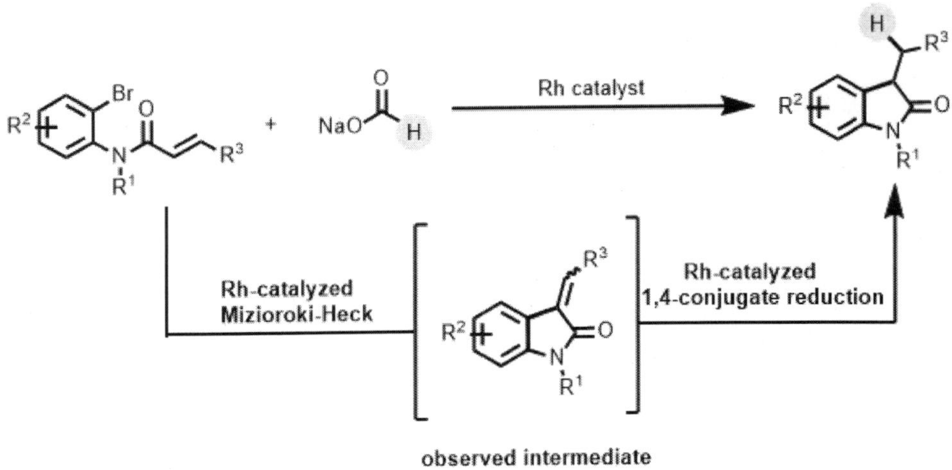

Scheme 11. Rh(I)-catalyzed domino Heck/1,4-hydride addition

Lautens et. al. presented[17] a domino method for the synthesis of oxindole-containing compounds starting with straightforward ortho-bromoaniline-derived acrylamides.A benzylidene oxindole intermediate is initially formed as part of this overall reaction. The creation of a rhodium hydride species is made possible by the use of sodium formate, and the subsequent addition of 1,4-hydride results in the final oxindole products.

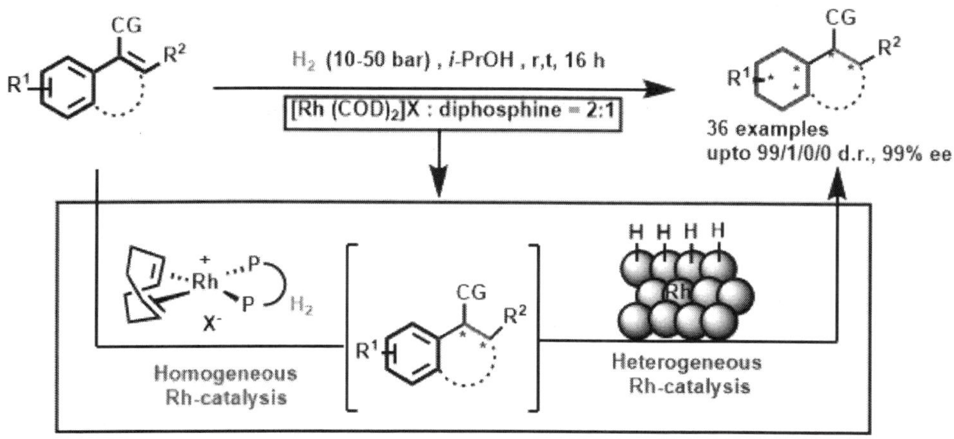

Scheme 12. Asymmetric saturation of vinylarenes with combination of homogeneous and heterogeneous Rhodium Catalysis

Dr. Anderson and his co-workers disclosed a study[18] describing the usage of diphosphine ligands along with common rhodium precursors for the cooperative homogeneous and heterogeneous, asymmetric complete hydrogenation of vinylarenes.This protocol provides a very adaptable and conceptually novel method based on a dual system: an irreversible binding between diphosphine ligands and rhodium, a very effective ligand-accelerated asymmetric hydrogenation of an olefin, and the *in situ* aggregation of nonligated rhodium metal into a very active nanoparticle for the hydrogenation of the aromatic ring, which leads to the simple construction of chiral cyclohexane scaffolds.

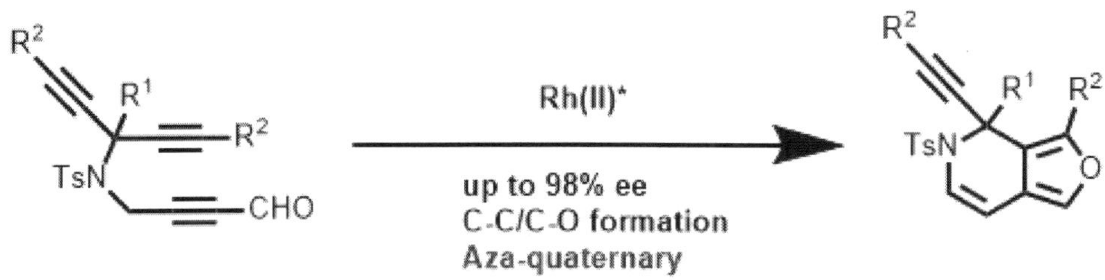

Scheme 13. Enantioselective Rh(II)-catalyzed desymmetric cycloisomerization

In bioactive compounds and natural products, alkynyl segments connected to carbon stereocenters are frequently present, and these molecules are important intermediates in synthetic chemistry.[19] Although useful, creation of alkynyl-containing tetrasubstituted carbon stereocenters bymeans of the desymmetric cyclization process is scarcely reported in literature. Dr. Zhu and his colleagues have reported[20] the first asymmetric intramolecularcycloisomerization of 1,6-diynes catalyzed by dirhodium(II)-complex. This methodology leads to highly functionalized furanfuseddihydropiperidine derivatives with simultaneous formationof an alkynyl incorporated aza-quaternary carbonstereocenter. High efficiency, good enantioselectivity, and atom economy are the important features of this technique.

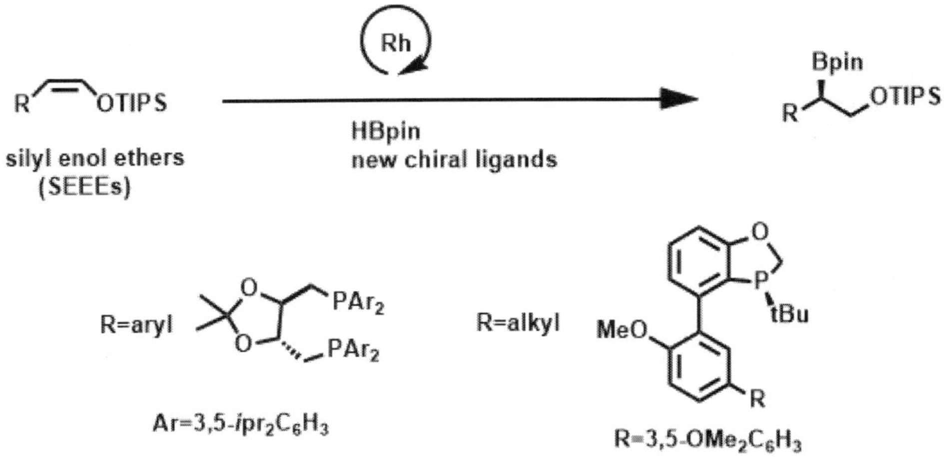

Scheme 14. Rh(I)-catalyzed hydroboration of silyl enol ethers

Dr. Zhao and his group have reported[21] Rh(I)-catalyzed asymmetrichydroboration of silyl enol ethers using two new phosphineligands. alkyl- andaryl-substituted silyl enol ethers both underwent the reaction and furnished high ee's. A gram-scale reaction and the changes in the borylether product serve as examples of the usefulness of this procedure. The representative scheme has been delineated in scheme 14.

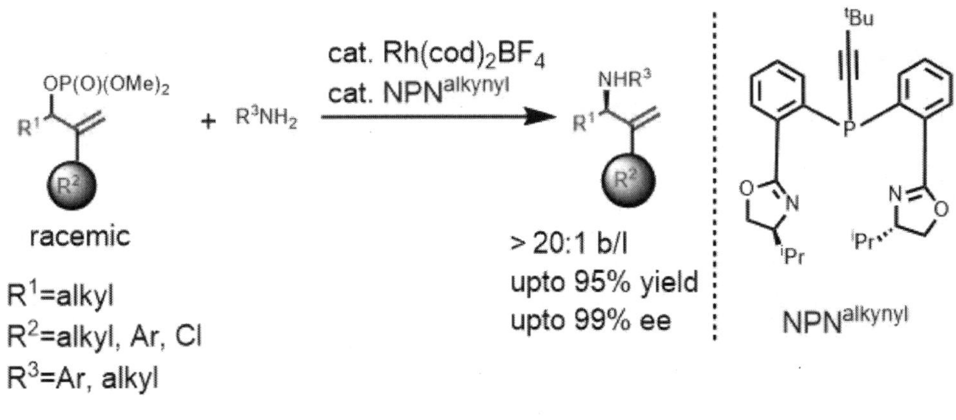

Scheme 15. Rhodium-catalyzed Regio- and Enantioselective Allylic Amination of racemic 1,2-Disubstituted Allylic Phosphates

In response to an examination of the monosubstituted allylrhodium-NPN complex, Li and his associates created and manufactured[22] brand-new bis(oxazoline) phosphine ligands with alkynyl functionalities on the P atom. These ligands could be employed to carry out Rh-catalyzed regio- and enantioselective allylic amination processesof challenging1,2-disubstituted allylic precursors having bulky 2-R′ groups. Chiral 1,2-disubstituted allylic amines with >20:1 branched/linear ratio and 99% ee were produced in up to 95% of the yields.

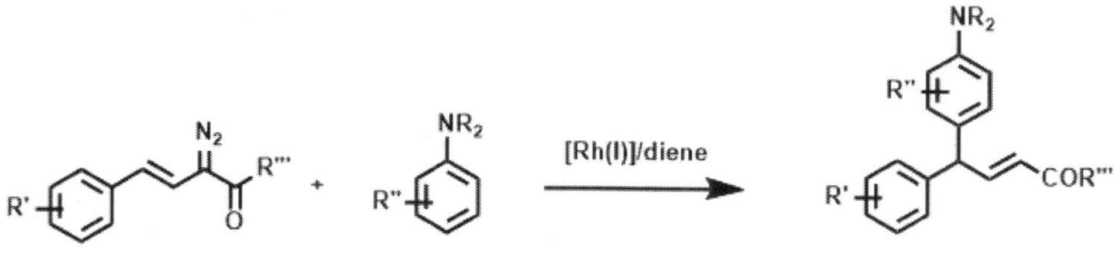

Scheme 16. Rh(I) carbene directed enantioselective C-H insertion

Xu and his group divulged23 the first example of the regiospecific and direct enantioselective C(sp2)-H functionalization of aniline derivatives with arylvinyldiazoacetates enabled by Rh(I)-diene catalysts. The catalysts demonstrate excellent performance in controlling the vinylogous reactivity of arylvinylcarbenoids to achieve enantioselective variants of this chemistry. Reaction proceeded in high yields and is compatible with and a wide range of functional groups, affording a variety of chiral γ,γ-gemdiaryl substituted , with excellent enantioselectivities under simple and mild conditions. Provides access to β-unsaturated esters.

Jiang and his collaborators disclosed[24]a Rh(I)-catalyzed carbene transfer/carbonylation/cyclization cascade reaction for the facile synthesis of fully substituted arynes with high potential for the synthesis of organic photovoltaic materials. The steric and hyperconjugative effects of silyl substituents on alkynes are important in the cascade process as they prevent oxidation of carbenoid intermediates. They are also currently conducting additional studies using this strategy to construct a variety of structurally complex aligners for organic light-emitting diodes (OLEDs).

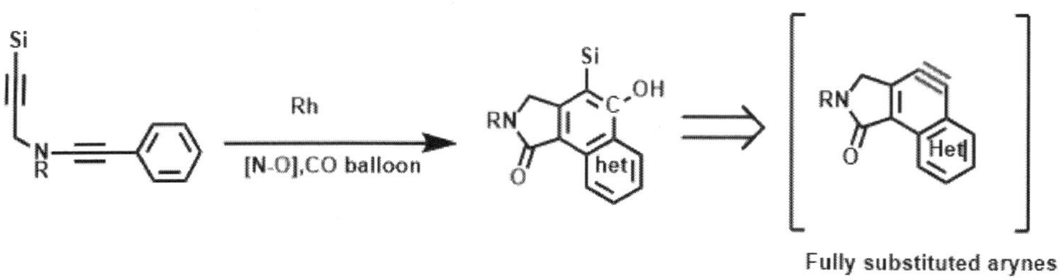

Scheme 17. Rh(I) catalyzed carbene migration

CONCLUSION

It is evident from the reactions described above that Rhodium can catalyze a diverse array of reactions with its different oxidation states *viz.* Rh(I), Rh(II), Rh(III). Among hundreds of Rhodium catalyzed reactions that have been reported in the literature recently, only few fascinating reactions have been chosen for this book chapter. A comprehensive review will be published later.

REFERENCES

[1] Fagnou, K.; Mark Lautens, M. Rhodium-catalyzed carbon-carbon bond forming reactions of organometallic compounds *Chem. Rev.* **2003**, *103*, 169-196.

[2] Greenwood, N. S.; Champlin, A. T.; Ellman, J. A. *J. Am. Chem. Soc.*2022, *144*, 17808 17814.

[3] Munnuri, S.; Falck, J. R. Directed, Remote Dirhodium C(sp3)-H Functionalization, Desaturative Annulation, and Desaturation *J. Am. Chem. Soc.* **2022**, *144*, 17989 17998.

[4] Tanner, D. Chiral Aziridines Their Synthesis and Use in Stereoselective Transformations. *Angew. Chem., Int. Ed.***1994**, *33*, 599 619.

[5] a) McCoull, W.; Davis, F. A. Recent Synthetic Applications of Chiral Aziridines. Synthesis **2000**, *2000*, 1347 1365. b) Hu, X. E. Nucleophilic Ring Opening of Aziridines. *Tetrahedron***2004**, *60*, 2701 2743. c) Singh, G. S.; D'hooghe, M.; De Kimpe, N. Synthesis and Reactivity of C-Heteroatom-Substituted Aziridines. *Chem. Rev.***2007**, *107*, 2080 2135.

[6] a) Forbeck, E. M.; Evans, C. D.; Gilleran, J. A.; Li, P.; Joullié, M. M. A Regio- and Stereoselective Approach to Quaternary Centers from Chiral Trisubstituted Aziridines. J. Am. Chem. Soc. **2007**, *129*, 14463 14469. b) Tanaka, H.; Sawayama, A. M.; Wandless, T. J. Enantioselective Total Synthesis of Ustiloxin D. *J. Am. Chem. Soc.***2003**, *125*, 6864 6865.

[7] Boquet, V.; Nasrallah, A.; Dana, A. L.; Brunard, E.; Di Chenna, P. H.; J. Duran, F. J.; Retailleau, P.; Darses, B.; Sircoglou, M.; Dauban, P.Rhodium(II)-Catalyzed Enantioselective Intermolecular Aziridination of Alkenes *J. Am. Chem. Soc.***2022**, *144*, 17156 17164.

[8] (a) Lawrence, S. A. Amines: Synthesis, Properties and Application; Cambridge University: Cambridge, UK, 2004. (b) Lamture, J. B. Aniline and Its Analogs; Notion Press: Chennai, India, 2018. (c) Roughley, S. D.; Jordan, A. M. The Medicinal Chemist's Toolbox: an Analysis of Reactions Used in the Pursuit of Drug Candidates. *J. Med. Chem.***2011**, *54*, 3451 3479. (d) Kienle, M.; Reddy Dubbaka, S.; Brade, K.; Knochel, P. Modern Amination Reactions. *Eur. J. Org. Chem.***2007**, 2007, 4166 4176.

[9] Chen, K.; Kang, Q.-K.; Li, Y.; Wu, W.-Q.; Zhu, H.; Shi, H. Catalytic Amination of Phenols with Amines. *J. Am. Chem. Soc.***2022**, *144*, 1144 1151.

[10] (a) Wehn, P. M.; Du Bois, J. Enantioselective Synthesis of the Bromopyrrole Alkaloids Manzacidin A and C by Stereospecific C H Bond Oxidation. *J. Am. Chem. Soc.***2002**, *124*, 12950 12951. (b) Kang, S. H.; Kang, S. Y.; Lee, H.-S.; Buglass, A. J. Total Synthesis of Natural tert-Alkylamino Hydroxy Carboxylic Acids. *Chem. Rev.***2005**, *105*, 4537 4558. (c) Beaulieu, P. L.; Lavallée, P.; Abraham; Anderson, P. C.; Boucher, C.; Bousquet, Y.; Duceppe, J.-S.; Gillard, J.; Gorys, V.; Grand-Maître, C.; Grenier, L.; Guindon, Y.; Guse, I.; Plamondon, L.; Soucy, F.; Valois, S.; Wernic, D.; Yoakim, C. Practical, Stereoselective Synthesis of Palinavir, a Potent HIV Protease Inhibitor. *J. Org. Chem.* **1997**, *62*, 3440 3448. (d) De Clercq, E. Antivirals: Past, present and future. *Biochem. Pharmacol.* **2013**, *85*, 727 744. (e) Chen, S.; Wang, Y.; Zhou, W.; Li, S.; Peng, J.; Shi, Z.; Hu, J.; Liu, Y.-C.; Ding, H.; Lin, Y.; Li, L.; Cheng, S.; Liu, J.; Lu, T.; Jiang, H.; Liu, B.; Zheng, M.; Luo, C. Identifying Novel Selective Non- Nucleoside DNA Methyltransferase

1 Inhibitors through Docking-Based Virtual Screening. *J. Med. Chem.* **2014**, *57*, 9028 9041. (f) Kelly, P. M.; Bright, S. A.; Fayne, D.; Pollock, J. K.; Zisterer, D. M.; Williams, D. C.; Meegan, M. J. Synthesis, antiproliferative and pro-apoptotic activity of 2-phenylindoles. *Bioorg. Med. Chem.* **2016**, *24*, 4075 4099. (g) Hems, W. P.; Groarke, M.; Zanotti-Gerosa, A.; Grasa, G. A. [(Bisphosphine) Ru(II) Diamine] Complexes in Asymmetric Hydrogenation: Expanding the Scope of the Diamine Ligand. *Acc. Chem. Res.* **2007**, *40*, 1340 1347. (h) Kizirian, J.-C. Chiral Tertiary Diamines in Asymmetric Synthesis. *Chem. Rev.* **2008**, *108*, 140 205.

[11] Yang, B.; Liu, X.; Yu, A.; Yang, Q.; Wang, Y. Rhodium(II)-Catalyzed Allylic 1,3-Diamination. *ACS Catal.* **2022**, *12*, 13411 13417.

[12] Zhao, W.; Chen, K.-Z.; Li, A.-Z.; Li, B.-J. Remote Stereocenter through Amide-Directed, Rhodium-Catalyzed Enantioselective Hydroboration of Unactivated Internal Alkenes. *J. Am. Chem. Soc.* **2022**, 144, 13071 13078.

[13] Ma, C.; Li, C.; Bai, J.; Xiao, J.; Zhai, Y.; Guo, Y.; Ma, S. Rhodium-Catalyzed Intermolecular Stereoselective Allylation of Indoles with Allenes. *ACS Catal.* **2022**, *12*, 10141 10146.

[14] Zhang, B.; Guo, X.; Tao, L.; Li, R.; Lin, Z.; Zhao, W. Rhodium-Catalyzed Regioselective and Chemoselective Deoxygenative Reduction of 1,3-Diketones. *ACS Catal.* **2022**, *12*, 4640 4647.

[15] Schiwek, C. H.; Jandl, C.; Bach, T. All-cis Saturated 2,5-Diketopiperazines by a Diastereoselective Rhodium-Catalyzed Arene Hydrogenation. *ACS Catal.* **2022**, *12*, 3628 3633.

[16] Ye, B.; Yao, J.; Wu, C.; Zhu, H.; Yao, W.; Jin, L.; Dou, X. Rhodium-Catalyzed Asymmetric Conjugate Pyridylation with Pyridylboronic Acids. *ACS Catal.* **2022**, 12, 2434 2440.

[17] Larin, E. M.; Masson-Makdissi, J.; Jang, Y. J.; Lautens, M. Rhodium(I)-Catalyzed Formate-Mediated Domino Heck/1,4-Hydride Addition toward Oxindoles. *ACS Catal.* **2022**, *12*, 12744 12749.

[18] Wu, H.; Yang, J.; Peters, B. B. C.; Massaro, L.; Zheng, J.; Andersson, P. G. Asymmetric Full Saturation of Vinylarenes with Cooperative Homogeneous and Heterogeneous Rhodium Catalysis. *J. Am. Chem. Soc.* **2021**, *143*, 20377 20383.

[19] a) Corbett, J. W.; Ko, S. S.; Rodgers, J. D.; Gearhart, L. A.; Magnus, N. A.; Bacheler, L. T.; Diamond, S.; Jeffrey, S.; Klabe, R. M.; Cordova, B. C.; Garber, S.; Logue, K.; Trainor, G. L.; Anderson, P. S.; Erickson-Viitanen, S. K. Inhibition of Clinically Relevant Mutant Variants of HIV-1 by Quinazolinone Non-Nucleoside Reverse Transcriptase Inhibitors. *J. Med. Chem.* **2000**, *43*, 2019. b) Trost, B. M.; Li, C.-J. Modern Alkyne Chemistry: Catalytic and Atom-Economic Transformations; Wiley-VCH: New York, 2014.

[20] Wu, R.; Lu, J.; Cao, T.; Ma, J.; Chen, K.; Shifa Zhu, S. Enantioselective Rh(II)-Catalyzed Desymmetric Cycloisomerization of Diynes: Constructing Furan-Fused Dihydropiperidines with an Alkyne-Substituted Aza-Quaternary Stereocenter. *J. Am. Chem. Soc.* **2021**, *143*, 14916 14925.

[21] Dong, W.; Xu, X.; Ma, H.; Lei, Y.; Lin, Z.; Zhao, W. Enantioselective Rh-Catalyzed Hydroboration of Silyl Enol Ethers. *J. Am. Chem. Soc.* **2021**, *143*, 10902 10909.

[22] Xu, W.-B.; Sun, M.; Shu, M.; Li, C. Rhodium-Catalyzed Regio- and Enantioselective Allylic Amination of Racemic 1,2-Disubstituted Allylic Phosphates. *J. Am. Chem. Soc.* **2021**, *143*, 8255 8260.

[23] Dong-Xing Zhu, Hui Xia, Jian-Guo Liu, Lung Wa Chung, and Ming-Hua Xu Regiospecific and Enantioselective Arylvinylcarbene Insertion of a C H Bond of Aniline Derivatives Enabled by a Rh(I)-Diene Catalyst. *J. Am. Chem. Soc.* **2021**, 143, 2608 2619.

[24] Zhu, G.; Gao, W.-C.; Jiang, X. Rh(I)-Catalyzed Carbene Migration/Carbonylation/Cyclization: Straightforward Construction of Fully Substituted Aryne Precursors *J. Am. Chem. Soc.* **2021**, *143*, 3, 1334–1340.

Index

2D Perovskites 35
3D Perovskites 32

A

Acetone 3
Activation energy 51
Alkaline 119
Ambipolar currents 87
Anilines 147
Anisotropic 111
Anton supercomputers 99
Aqueous solution 55
Arc Discharge Method 69
Atomic layer deposition (ALD) 12
Atomic radius 48
Atomistic 99

B

Binary fluids 23
Biocompatibility 67
Bioimaging 67
Biomolecules 97
Bio-preservation 28
Biosensors 11
Bromine (Br) 123
Bulk material 69
Bulky groups 37
Butyl ammonium 35
B-X bond 122
Byoscillations 26

C

Cancer cells 67
Carbon dots 67
Carbon fiber 69
Carbon nanotubes 67
Carbon sooth 69
Carl Zeiss 12
Catalytic systems 80
Cellulose nanocrystals 100
Cesium chloride (CsCl) 119
Chemical Oxidation Method 69
Chemical vapour deposition (CVD) 12
Cheung's method 4
Cooling process 26
Cotectic 27
Crowding effect 98
Crystallization 23
Crystallographic 13
Crystal symmetry 119
Cubic 122
Cytoplasmic volume 98

D

Dichalcogenides 48
Dichroic ration (DR) 112
Dihydrate crystals 25
Dihydrogen 79
Dye solution 56

E

Earth metal 119
Ecosystem 55
Electric quadrupole 18
Electromagnetic radiation 24
Electronic devices 2
Electron microscope 49
Energy conversions 67
Eutectic 25
Eutectic mixture 25

F

Fabrication 109
FA-dehydrogenation 80
Ferroic 41
Ferromagnetic 43
Fibre-optic sensors 24
Film transfer method (FTM) 109
Formic Acid 79
Fresnel reflection 23
Fuel cell 80
Fuel cells 79

G

Gelation 99
Gene therapy 100
Genotoxic 55
Graphene oxide 69
Green's function 88

H

Halogen 117
H-atoms 81
Health problems 55
Hetero atom 68
Heterojunction 58
Heterojunction Photocatalysts 59
Hexagonal 13
Hole transporting material (HTL) 103
Homo energy 105
Homogeneous Condition 82
Hydrogels 99
Hydrogenation 80
Hydrogen economy 79
Hydrogen Production 80
Hydrohalite 26
Hydrophobic 37
Hydrothermal method 52
Hydrothermal Method 70
Hydrothermal reaction 12
Hydrothermal synthesis 48

I

Interfacial layers 3

K

K-isoelectronic 17

L

Laser Ablation Method 69
Lead (Pb) 119
Liquid nitrogen 24
Logarithmic scale 7
Lyophilization 59

M

Macromolecular Crowding 99
Magnetic dipole 18
Magnetoelectric 41
Magnetoelectric (ME) 41
Metallic silver 50
Methanol 3
Methylammonium 31
Microfiltration 56
Microspheres 60

Microwave Assisted Method 71
Mineralogy 28
Molecules 55, 80
Molybdenum 47, 49
Molybdenum atom 48
Morphology 11
Multiferroic 41
Multiferroics 41
Multifunctional devices 41

N

Nanomaterials 67, 98
Nanoparticles 80
Nanoplates 58
Nanotechnology 99
N-butylamine (BA) 32
Neurotoxic 55
Nickel 111
Norde method 3, 5
Nuclei 13
Nucleic acids 97
Nucleophiles 146

O

Optoelectronic device 52
Optoelectronics 67
Organic dye 8
Orthorhombic structures 122
Oscillators 41
Oxalic acid 12
Oxalic acid 13
Oxygen pressure 42

P

Palladium 80
Periodic table 81
Perovskite design 119
Perovskites 31
Perovskite solar cells (PSC) 103
Phenylethylamine (PEA) 32
Photocatalysis 56
Photodetector 110
Photodetectors 11
Photoelectrons 58
Photoelectrons migration 60
Photoluminescence 11
Photoreduction 56
Photovoltaic 106
Physicochemical methods 56
Physiological functions 55
Poisson's equation 104
Polyethylene glycol (PEG) 98
Poole-Frenkel 2
Prussian Blue 56
Pyrolytic Method 71
Pyronine-B 2

R

Raman spectrum 14
Rb-isoelectronic 17
Redox reaction 56
Refractive index 25
Rhodamine B 56, 57, 63
Rubrene 2

S

Scanning electron microscopy (SEM) 43
Schottky 6
Schottky diodes 8
Schrodinger equation 88
Si-based solar cells 37
Si-cells 32
Silver atoms 47

Slope factors 7

Sodium tungstate 12

Solar cells 32

Solvothermal Method 70

Sorbitol 71

Spectra 34

Spectrophotometer 12

Spectroscopy 47, 57

Spectrum illumination 105

SrTiO3 (STO) 44

Sulforhodamine-G 2

Supramolecular fibers 99

Synthetic dyes 55

Synthetic methods 68

Synthetic methods 68

T

Tedious 56

Teflon 12

Ternary solution 26

Thioacetamide 12

Trichloroethylene 3

Trigonal prismatic 48

Trihalide 115

Trimetallic systems 82

Tungsten disulphide 11

U

Ultrasonic method 69

Ultrasonic Method 70

UV irradiation 56

UV light 58

W

Water pollution 55

X

X-ray 34

x-ray diffraction (XRD) 42

XRD spectrum 49